CRYSTAL STRUCTURES AND CATION SITES OF THE ROCK-FORMING MINERALS

TITLES OF RELATED INTEREST

An atlas of optical transforms
G. Harburn *et al.*
Boninites
A. J. Crawford (ed.)
Carbonatites
K. Bell (ed.)
Cathodoluminescence of geological materials
D. Marshall
The dark side of the Earth
R. Muir Wood
Environmental chemistry
P. O'Neill
A first course in crystallography
A. Windle
Image interpretation in geology
S. A. Drury
The inaccessible Earth
G. C. Brown & A. E. Mussett
The interpretation of igneous rocks
K. G. Cox *et al.*
An introduction to X-ray spectrometry
K. L. Williams
Komatiites
N. T. Arndt & E. G. Nisbet (eds)
Mathematics in geology
J. Ferguson
Metamorphic geology
C. Gillen
Metamorphic processes
R. H. Vernon
Petrology of the igneous rocks
F. H. Hatch *et al.*
Petrology of the metamorphic rocks
R. Mason
A practical introduction to optical mineralogy
C. D. Gribble & A. J. Hall
Rheology of the Earth
G. Ranalli
Rutley's elements of mineralogy
H. H. Read
Simulating the Earth
J. Holloway & B. Wood
Statistical methods in geology
R. F. Cheeney
Volcanic successions : modern and ancient
R. A. F. Cas & J. V. Wright
The young Earth
E. G. Nisbet

CRYSTAL STRUCTURES AND CATION SITES OF THE ROCK-FORMING MINERALS

Joseph R. Smyth
Department of Geological Sciences, University of Colorado
and
David L. Bish
Los Alamos National Laboratory

Boston
ALLEN & UNWIN
London Sydney Wellington

This book was typeset, proofed and passed for press by the authors.

Allen & Unwin, Inc.,
8 Winchester Place, Winchester, Mass. 01890, USA
the U.S. company of
Unwin Hyman Ltd
PO Box 18, Park Lane, Hemel Hempstead, Herts HP2 4TE, UK
40 Museum Street, London WC1A 1LU, UK
37/39 Queen Elizabeth Street, London SE1 2QB, UK

Allen & Unwin (Australia) Ltd,
8 Napier Street, North Sydney, NSW 2060, Australia
Allen & Unwin (New Zealand) Ltd in association with the Port Nicholson Press Ltd,
60 Cambridge Terrace, Wellington, New Zealand

First published in 1988

Library of Congress Cataloging-in-Publication Data

Smyth, J. R.
Crystal structures and cation sites of the rock-forming minerals.
Bibliography: p.
Includes index.
1. Rock-forming minerals. 2. Crystals. 3. Cations.
I. Bish, David L. II. Title.
QE397S65 1987 549′.18 87-14526
ISBN 0-04-445004-4 (alk paper)

British Library Cataloguing in Publication Data

Smyth, Joseph R.
Crystal structures and cation sites of the rock-forming minerals.
1. Rock-forming minerals
I. Title II. Bish, David L.
549′.6 QE397
ISBN 0-04-445004-4

Printed in Great Britain by
Biddles Limited, Guildford, Surrey

CONTENTS

For Tamsin and Karen

ACKNOWLEDGEMENT

This work was supported in part by the U.S. Department of Energy, Office of Basic Energy Sciences, through several grants to Los Alamos National Laboratory which is operated by The University of California under contract number W-7405-ENG-6 The authors particularly thank Dr. George Kolstadt (OBES - Chemistry, Earth, and Life Sciences) and Dr. Ryszard Gajewski (OBES - Advanced Energy Projects) for generous support of the project. The authors thank Drs. Y. Ohashi (ARCO, Plano, TX), R. X. Fischer (Johannes Gutenberg Universitaet, Mainz), and L. W. Finger (Carnegie Institution, Washington, DC) for providing computer codes and discussions and Drs. George Zweig, Klaus Lackner, and Wes Myers (Los Alamos National Laboratory) for discussions, support, and encouragement throughout the project. Theoretical Division Office of Los Alamos National Laboratory is also thanked for its support, and Tamsin C. McCormick is gratefully acknowledged for tireless proofreading, technical assistance and moral support.

INTRODUCTION

Over the past two decades, with the advent of automated x-ray and neutron single-crystal diffractometers, there has been a major improvement in the precision with which atom positions in minerals are known. Shannon and Prewitt (1969, 1970), Whittaker and Muntus (1970), and Shannon (1976) have compiled crystal structure data for synthetic compounds and minerals in order to estimate effective ionic radii. These compilations and estimates have proven immensely useful to geochemists, mineralogists, and petrologists in understanding the substitution behavior and distribution of elements in natural systems, eg. Onuma et al. (1968), Jensen (1973), Philpotts (1978). Whereas Bragg et al. (1965) and Zoltai and Stout (1984) have compiled descriptions of mineral structures, and Wyckoff (1963) has compiled atom location data for most inorganic structures, there has never been a compilation of data on the nature of cation sites in minerals.

Robie et al. (1978) compiled thermodynamic data for many of the rock-forming minerals and oxides. For some of these compounds there have been more recent and accurate cell determinations, so that improved data on molar volume and density are available. Further, thermodynamic data compilations do not include information on atomic environments in these compounds.

We have undertaken a compilation of recent data on crystal structures for a large group of the common minerals. From atom positional and cell data, we have calculated nearest-neighbor distances, coordination numbers, volumes of coordination polyhedra, distortion indices, and electrostatic energies in a consistent fashion. The objective in this work is to make the recent improvements in crystal structure data available to a larger group of petrologists and geochemists seeking to understand the chemical behavior of these minerals in natural systems.

In order to reduce this to a manageable task we have had to make some rather arbitrary decisions in selecting and grouping the data. First, we have limited the the group to the oxygen and halide minerals with the understanding that ionic radii have at least some relevance to these structures. This has led to the exclusion of the sulfides from the current compilation. In selecting structures, we have endeavored to choose ordered end-members whenever possible so that cation site data will be more easily interpretable. In order to document atomic environments in standard thermodynamic states we have included a large number of simple oxide minerals. This has led to the inclusion of some less-than-common minerals in this group, but otherwise we have included only the more common minerals of igneous, metamorphic, and sedimentary rocks. Finally, in order to facilitate comparisons, we have grouped together data from isomorphous structures, and in a few cases, polymorphous structures. This has led to a few instances of duplication which we feel are justified in order to allow comparisons.

Unit Cell Tables

We have organized the structure data into those pertaining to unit cells and those pertaining to specific sites. In addition, we have summarized the site data, grouped them according to coordination number, and listed them by mean distance in Chapter 11. Within the mineral groups, unit cell tables consist of formula, formula weight, calculated density, molar volume, Z, crystal system, class, and space group, cell parameters, and reference. In general, the formula is that given in the reference, except that we have omitted elements constituting less than 1.0 weight percent of the mineral. In a few instances we have recalculated formulas to the same number of oxygens for comparison across an isomorphous series. The formula weight, density, and molar volume are our calculation from the stated formula and cell. Z is the number of formula units per cell. The reference is not repeated in the site tables, but site data are presented sequentially in the same order permitting unambiguous citation.

Site Data Tables

Similar sites in isomorphous series are grouped together to facilitate comparisons and show variability of analogous features across the series. The tabulated data consist of a site name, coordination number (C.N.), occupants, point symmetry, Wyckoff notation, fractional coordinates, nearest neighbor distances, mean and standard deviation of distances, polyhedral volume, quadratic elongation, variance of central angle, electrostatic site energy, and a model charge. The coordination number is the number of nearest anion neighbors. The occupant is that inferred from the formula or stated in the reference. In a few instances, for partially occupied sites, a total site occupancy is given. Tetrahedral Al-Si occupancies for some of the zeolites were calculated from the mean T-O distance when site occupancies were not reported. The point symmetry and Wyckoff notation are those for the site (Hahn, 1983). Fractional coordinates for the site are included to avoid any ambiguities in site nomenclature that may arise and to show variability across the series. Individual nearest neighbor distances are given throughout with a major exception being those for the framework silicate structures (Chapter 8). With the low symmetries of many of these structures, it was found very difficult to present these in a way that would be both concise and meaningful. Also, we have omitted the cavity geometries for the zeolites as these are documented elsewhere (Mortier, 1982).

The mean distance is our calculated average of the given distances. The σ is the standard deviation of the distances. It is thus an estimate of the distortion of the site, not an estimate of the error in the determination. Errors are regrettably not given because this would have more than doubled the size of the data base, greatly complicating the handling of the data. The polyhedral volume (Poly. Vol.), quadratic elongation (Q.E.), and angle variance (Ang.Var.) were calculated with a slightly modified version of the program VOLCAL (L.W. Finger, personal communication). The units of polyhedral volume are cubic Angstroms. Quadratic elongation as defined by Robinson et al. (1971) is unitless, and the units of the variance of the central polyhedral angle are degrees squared. These two quantities are defined only for octahedra and tetrahedra. The electrostatic site energy was calculated with the program ELEN (Y. Ohashi, personal communication). Calculations were performed on a VAX 11/750 computer using double-precision arithmetic. High symmetry structures were reduced to triclinic symmetry. The units are kcal/mole and are the amount of electrostatic energy derived by placing 1 mole of cations of the stated charge into the site, assuming a purely point charge model. This energy is recalculated as electron volts (eV) in Chapter 11. The model charge is that used for the electrostatic calculation, however this may be omitted in instances where it is an integer unambiguously inferred from site occupants. These energies are included for qualitative comparisons among sites and should not be used for quantitative calculations because they exclude repulsive forces entirely. In addition, partial charges do not accurately model the effects of disorder.

Systematic Variation of Site Parameters

With a data base of this size, it is relatively straightforward to examine correlations between various site parameters. Such correlations help explain the nature of variations seen from structure to structure. Two particularly useful correlations are between angle variance and quadratic elongation and between electrostatic potential and mean cation-anion distance for individual sites.

Distortions. There are several parameters that can be used as indicators of distortions or regularity of coordination polyhedra. For regular tetrahedra and octahedra, the angular distortions are conveniently indicated by the variance of the central angle (Ang.Var.). In addition to angular distortion is distance distortion, a convenient measure of which is the standard deviation of the distances (σ) which can be used to indicate distortion of cations of any coordination number. A factor called quadratic elongation (Q.E.) (Robinson et al., 1971) is also calculated for each octahedron and tetrahedron and is a convenient measure of both angular and distance distortions. Figure 0.1 is a plot of angle variance versus quadratic elongation for a large number of tetrahedra and shows that the two parameters are, of course, strongly correlated. There are a few instances in which the angle variance plots well below the trend, but no instances in which it plots above. This is likely due to the fact that in a few instances, such as for a tetrahedron on three-fold axis, the angles may be more strongly constrained by symmetry than the distances.

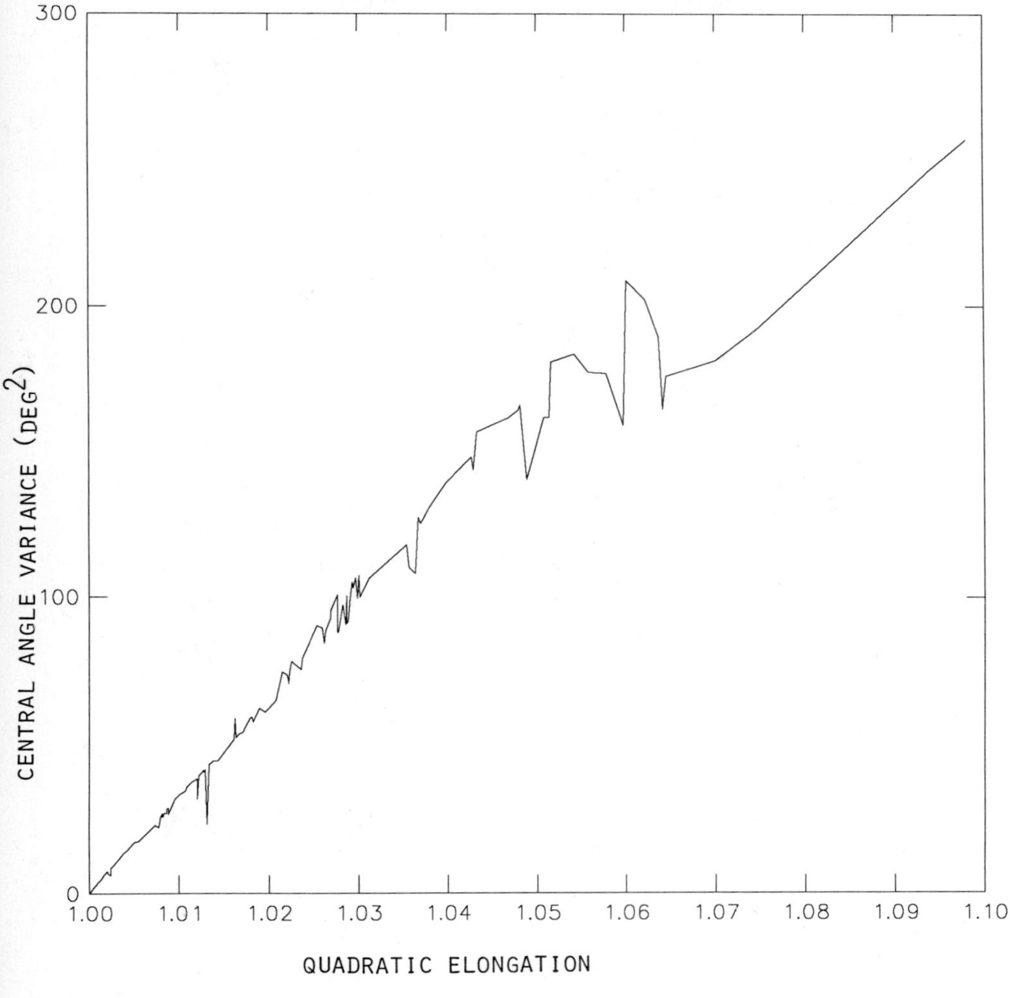

Figure 0.1. Plot of angle variance versus quadratic elongation for SiO_4 tetrahedra. Adjacent points are conencted to show not only the strong correlation between the two factors, but also that the angle variance occasionally falls below, but not above, the general trend (see text).

Electrostatic Energies. It is also of interest to examine the variation of electrostatic energy with distance. The electrostatic energy reported in the site tables is in kcal/mole of sites. In Chapter 11, these energies are converted to electron-volts and divided by the charge to give a potential in volts. The total electrostatic energy of the crystal would then be half the sum for all sites in the formula unit. The total electrostatic energies calculated for each of these mineral structures is adequate to allow full ionization of all species to their normal valences, if reasonable allowances for repulsion energies are included in Born-Haber calculations.

The electrostatic energy is by no means the total energy of the crystal. It specifically excludes nearest-neighbor repulsion energies which may be ten percent or more of total energy. In addition, it excludes any estimate of distortion energies of electron distributions (e.g. crystal field stabilization energies) and energies of thermal vibrations. The energies cannot be used to compute heats of formation or predict relative thermodynamic stability of various polymorphs because the energy differences between polymorphs is typically much smaller than the excluded terms. The energies are, however, useful and instructive for qualitative comparisons between sites. Further, much progress has been made in recent years on prediction of mineral structures based on electrostatic energies combined with simple to complex expressions for nearest-neighbor repulsion energies (e.g., Catlow et al., 1982, Price and Parker, 1984).

We have plotted electrostatic potential (eV/chg) versus mean cation-anion distance for some 700 sites (Figure 0.2) and observe a very strong correlation. This figure shows clearly that there is a systematic electrostatic contribution to the energy of the crystals. Further, we have preliminary indication that deviations from the observed trend are significant and potentially useful indicators of minor element substitutions.

Trace and Minor Element Substitutions

One of the major reasons for undertaking this compilation was to provide geochemists with a convenient and comprehensive source of information on mineral sites. We hope that it will be of use in understanding trace element and

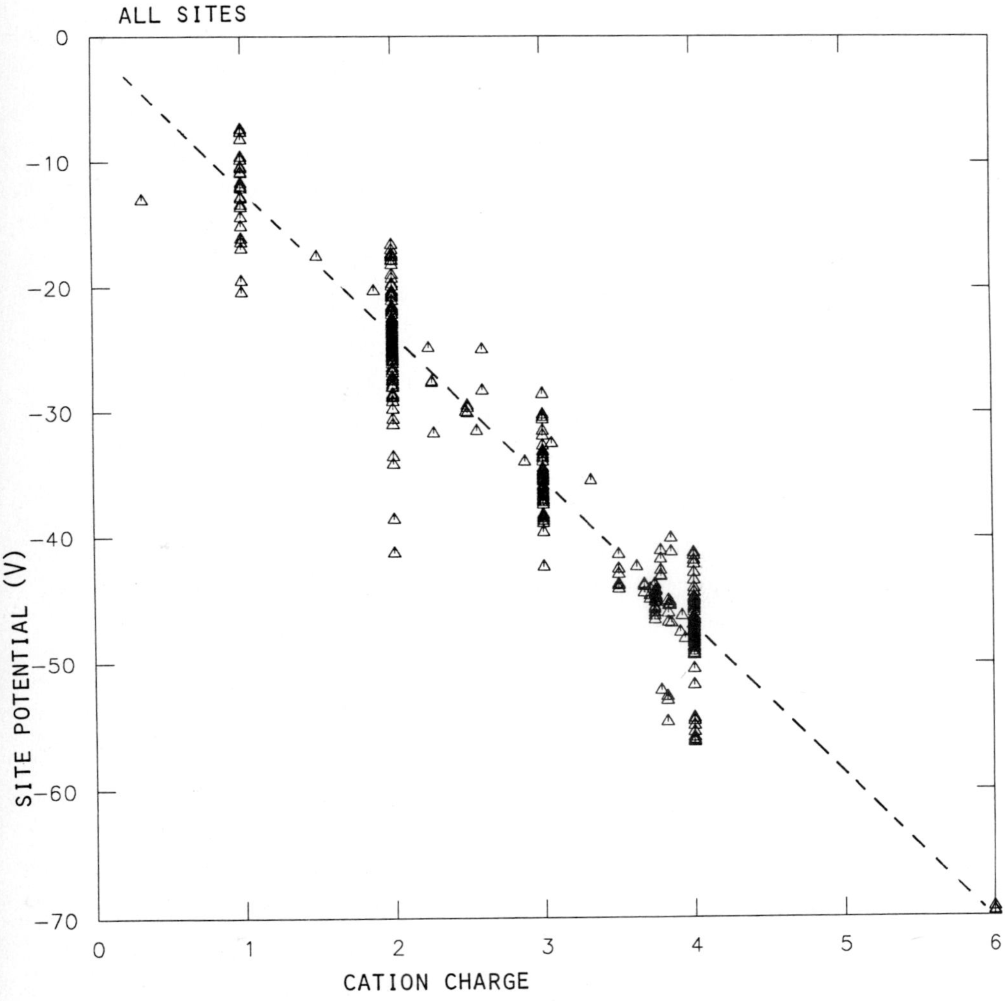

Figure 0.2. Plot of electrostatic site potential (V) versus cation charge. The plot shows the range of variation around a strong, nearly linear trend.

minor element distributions in geochemical systems. We have noted a few interesting correlations and general observations that are worth mentioning here. We hope that users of this volume will find many more.

The electrostatic energy and its variation with mean cation-anion distances may be a potentially useful indicator of minor element substitution sites. For example, of the two large cation sites in the epidote group (5.1), the A2 site has the much deeper electrostatic potential well despite its larger volume. It has, in fact the largest electrostatic energy per charge (potential) of any site surveyed with coordination number greater than 8 (Table 11.7). This accounts for the preference of trivalent rare earth elements for the A2 site over the A1 site in allanite and may explain why allanite has been observed to have distribution coefficients for rare earths relative to whole rocks of 1000 or more. It is relatively straightforward, then to identify potential sites for lanthanides and actinides from Tables 11.6 and 11.7.

We have also noted that sites tend to favor minor element substitutions that minimize distortions of the site or of the mineral as a whole, that is, of the other sites in the mineral as well. Recent geochemical studies have noted that the rhombohedral carbonates siderite and calcite both show large distribution coefficients for Mn relative to aqueous fluids (Ishikuni, 1984). In plotting distortion coefficients of the octahedral site versus mean distance in pure carbonates, we see a minimum in·the distortion, particularly in angle variance, near Mn. This is in marked contrast to most silicate octahedral sites which show a preference for smaller cations and may reflect a smaller effective radius for oxygen in the silicate octahedra relative to the carbonate octahedra. This may also be true for a broad range of octahedral sites in silicates and may account for the observation of Goldschmidt (1958) that in general mineral sites prefer smaller, rather than larger, cations. We note that for many silicates (e.g. olivine and orthopyroxene) the distortions of the octahedral sites decrease with radius at least down to the radius of Ni and that these structures have a strong preference for smaller cations in these sites.

The preference for smaller cations by silicate sites is certainly not applicable to sites other than regular octahedra. The X site in garnet is a fine example. It may contain Ca, Mn, Fe, or Mg and is highly symmetric with point symmetry 222. Its only measure of distortion, σ, does not vary strongly with cation radius, but the distortion of the Si site decreases strongly with increasing X-site radius. We would predict from these considerations then that if garnet crystallized as a liquidus phase, as it apparently does in the eclogite system, it would preferentially accept Ca, then Fe, then Mg, and thus possibly enrich residual liquids in Mg.

We have included these discussions on major and trace element distributions to encourage users of these tables to look for correlations between crystal structure parameters and element distributions in natural systems. It is our sincere hope in compiling these data that they will lead to greater understanding of geochemistry as well as mineralogy.

Chapter 1
SINGLE OXIDES

Table 1.1.1. Cuprite and Ag_2O Unit Cells.

End-Member	Cuprite	Ag_2O
Formula	Cu_2O	Ag_2O
Form.Wt.	143.079	231.739
Density	6.104	7.318
Mol.Volume	23.439	31.667
Z	2	2
Cryst.Sys.	Isometric	Isometric
Cryst.Class	$m3m$	$m3m$
Space Group	$Pn3m$	$Pn3m$
Cell Parameters		
a	4.2696	4.720
Vol.	77.833	105.15
Ref.	Borie (1974)	Borie (1974)

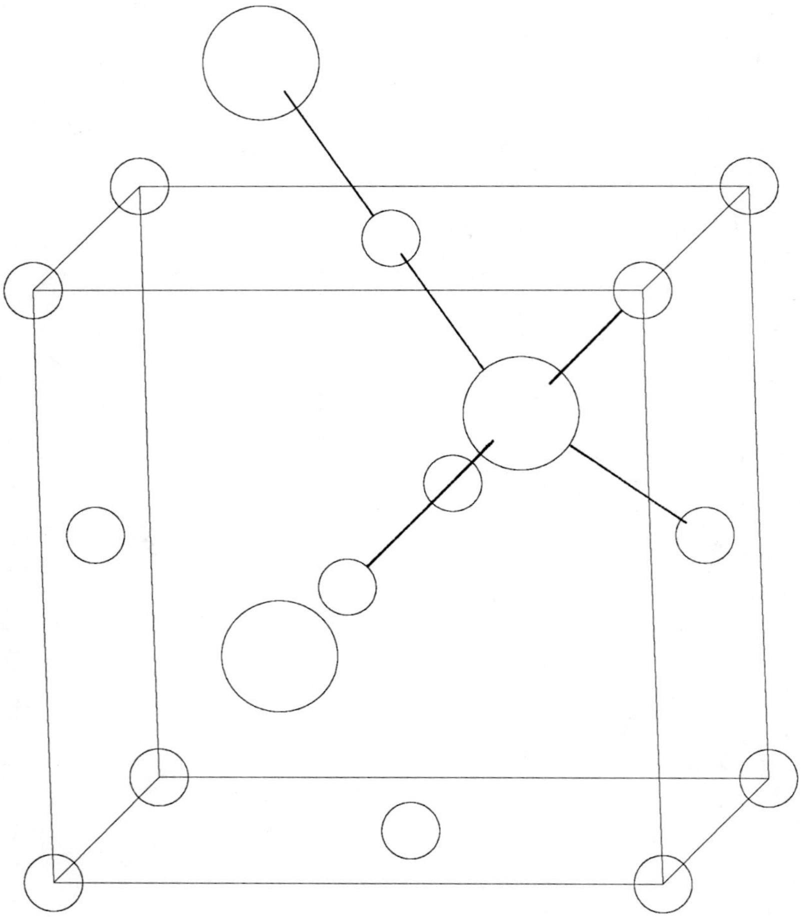

Figure 1.1. Cuprite, Cu_2O, perspective a-axis projection. Cu (small circle) is in linear two-coordination with point symemtry $\bar{3}m$. O is in perfectly regular tetrahedral coordination with point symmetry $\bar{3}m$. Synthetic Ag_2O also has this structure, but no other oxides do. The alkali metal oxides have the antifluorite structure but do not occur as minerals.

Table 1.1.2. Cuprite and Ag_2O Cation Sites.

End-Member	Cuprite	Ag_2O
C.N.	2	2
Occupant	Cu	Ag
Point Sym.	$\bar{3}m$	$\bar{3}m$
Wyckoff Not.	4b	4b
Frac.Coord.		
x	0	0
y	0	0
z	0	0
Distances		
O1(2)	1.849	2.044
Site Energy	-294.	-266.

Table 1.2.1. Periclase Group Unit Cells.

End-Member	Periclase	Wustite	Bunsenite	Manganosite	Lime
Formula	MgO	FeO	NiO	MnO	CaO
Form.Wt.	40.311	71.848	74.709	70.937	56.079
Density	3.585	5.956	6.850	5.365	3.346
Mol Volume	11.244	12.062	10.906	13.233	16.762
Z	4	4	4	4	4
Cryst.Sys.	Isometric	Isometric	Isometric	Isometric	Isometric
Cryst.Class	$m\,3m$	$m\,3m$	$m\,3m$	$m\,3m$	$m\,3m$
Space Group	$Fm\,3m$	$Fm\,3m$	$Fm\,3m$	$Fm\,3m$	$Fm\,3m$
Cell Parameters					
a	4.211	4.3108	4.1684	4.446	4.8105
Vol.	74.67	80.11	72.43	87.88	111.32
Ref.	Hazen (1976)	Foster & Welch(1956)	Wyckoff (1963)	Sasaki et al. (1980)	Wyckoff (1963)

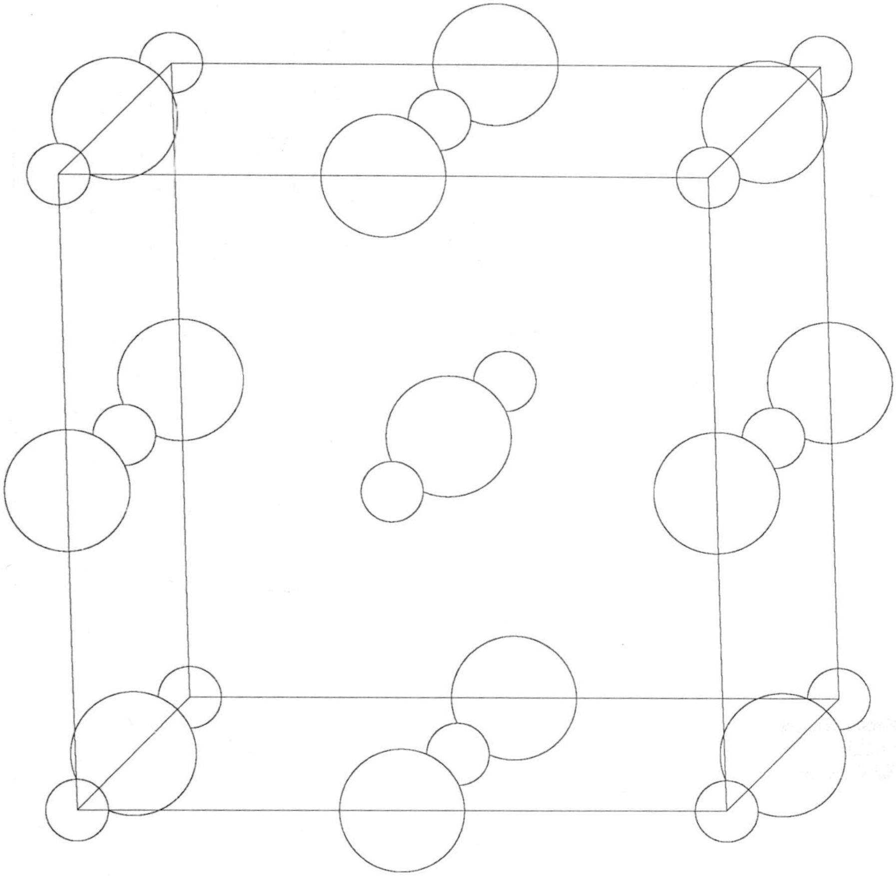

Figure 1.2. Periclase, MgO, perspective a-axis projection. Both Mg (small circle) and O (large circle) are in perfectly regular octahedral coordination with point symmetry $m3m$. This is the "rocksalt" structure of halite ($NaCl$) and sylvite (KCl). It is also the structure of many divalent metal oxides of which wustite (FeO), bunsenite (NiO), manganosite (MnO), and lime (CaO) occur as minerals. Galena (PbS) also has this structure.

Table 1.2.2. Periclase Group Octahedral Sites.

End-Member	Periclase	Wustite	Bunsenite	Manganosite	Lime
C.N.	6	6	6	6	6
Occupant	Mg	Fe	Ni	Mn	Ca
Point Sym.	$m3m$	$m3m$	$m3m$	$m3m$	$m3m$
Wyckoff Not.	4a	4a	4a	4a	4a
Frac.Coord.					
x	0	0	0	0	0
y	0	0	0	0	0
z	0	0	0	0	0
Distances					
O1(6)	2.105	2.155	2.084	2.223	2.405
Poly.Vol.	12.445	13.351	12.071	14.647	18.553
Q. E.	1.0000	1.0000	1.0000	1.0000	1.0000
Ang. Var.	0	0	0	0	0
Site Energy	-1102.	-1077.	-1114.	-1044.	-965.

Table 1.3.1. Zincite Group Unit Cells.

End-Member	Zincite	Bromellite
Formula	ZnO	BeO
Form.Wt.	81.369	25.012
Density	5.712	3.080
Mol Volume	14.246	8.122
Z	2	2
Cryst.Sys.	Hexagonal	Hexagonal
Cryst.Class	$6mm$	$6mm$
Space Group	$P\,6_3mc$	$P\,6_3mc$
Cell Parameters		
a	3.2427	2.6984
c	5.1948	4.2770
Vol.	47.306	26.970
Ref.	Sabine & Hogg (1969)	Sabine & Hogg (1969)

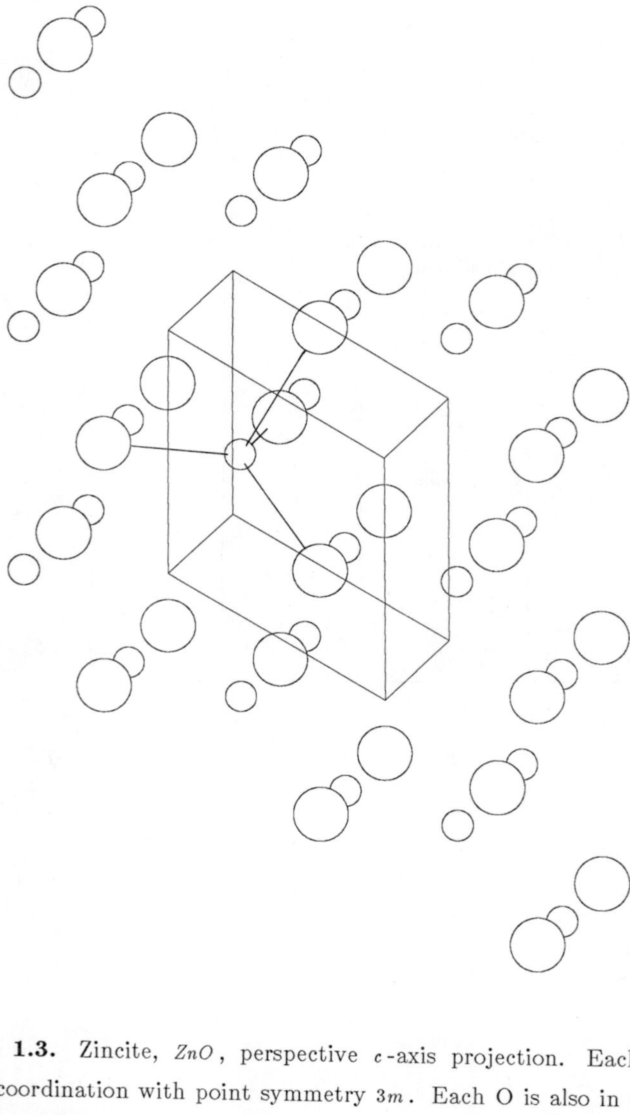

Figure 1.3. Zincite, ZnO, perspective c-axis projection. Each Zn is in tetrahedral coordination with point symmetry $3m$. Each O is also in tetrahedral coordination, but there are two distinct $Zn-O$ distances. This is sometimes known as the "wurtzite" structure of hexagonal ZnS and is also the structure of bromellite, BeO.

Table 1.3.2. Zincite Group Cation Sites.

End-Member	Zincite	Bromellite
C.N.	4	4
Occupant	Zn	Be
Point Sym.	3m	3m
Wyckoff Not.	2b	2b
Frac.Coord.		
x	1/3	1/3
y	2/3	2/3
z	0	0
Distances		
O1(1)	1.988	1.619
O2(3)	1.969	1.642
Mean	1.974	1.636
σ	.009	.011
Poly.Vol.	3.942	2.247
Q. E.	1.0006	1.0005
Ang. Var.	2.6	1.5
Site Energy	-1105.	-1333.

Table 1.4.1. Tenorite and Montroydite Unit Cells.

End-Member	Tenorite	Montroydite
Formula	CuO	HgO
Form.Wt.	79.539	216.589
Density	6.515	11.193
Mol Volume	12.209	19.350
Z	4	4
Cryst.Sys.	Monoclinic	Orthorhombic
Cryst.Class	$2/m$	mmm
Space Group	$C2/c$	$Pnma$
Cell Parameters		
a	4.6837	6.612
b	3.4226	5.520
c	5.1288	3.521
β	99.54	
Vol.	81.080	128.51
Ref.	Asbrink & Norrby (1970)	Aurivilius (1956)

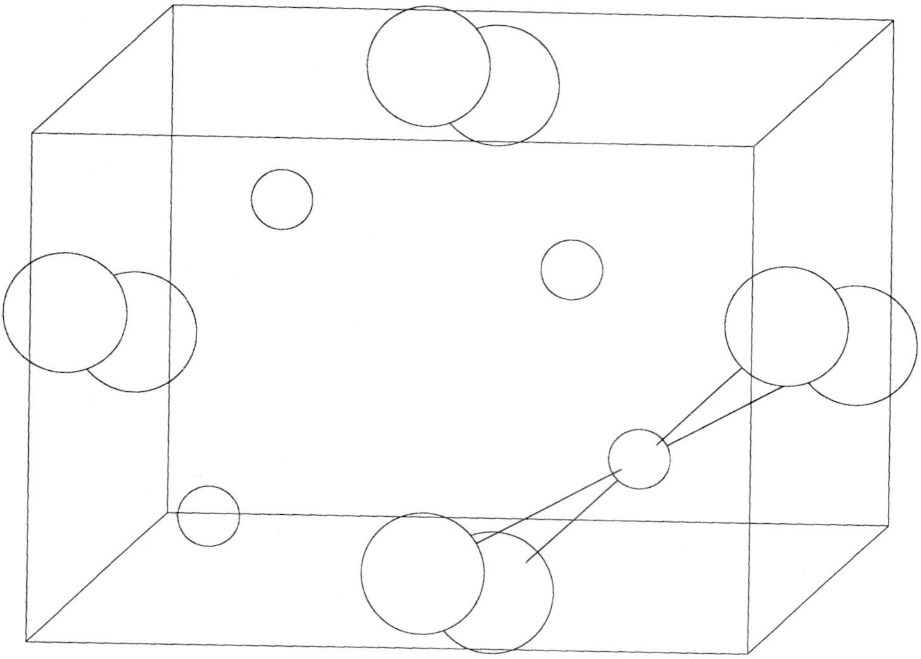

Figure 1.4a. Tenorite, CuO, perspective c-axis projection, a horizontal, b vertical. Each Cu (small circle) is in planar four-coordination with point symmetry $\bar{1}$, and each O (large circle) is in irregular four-coordination.

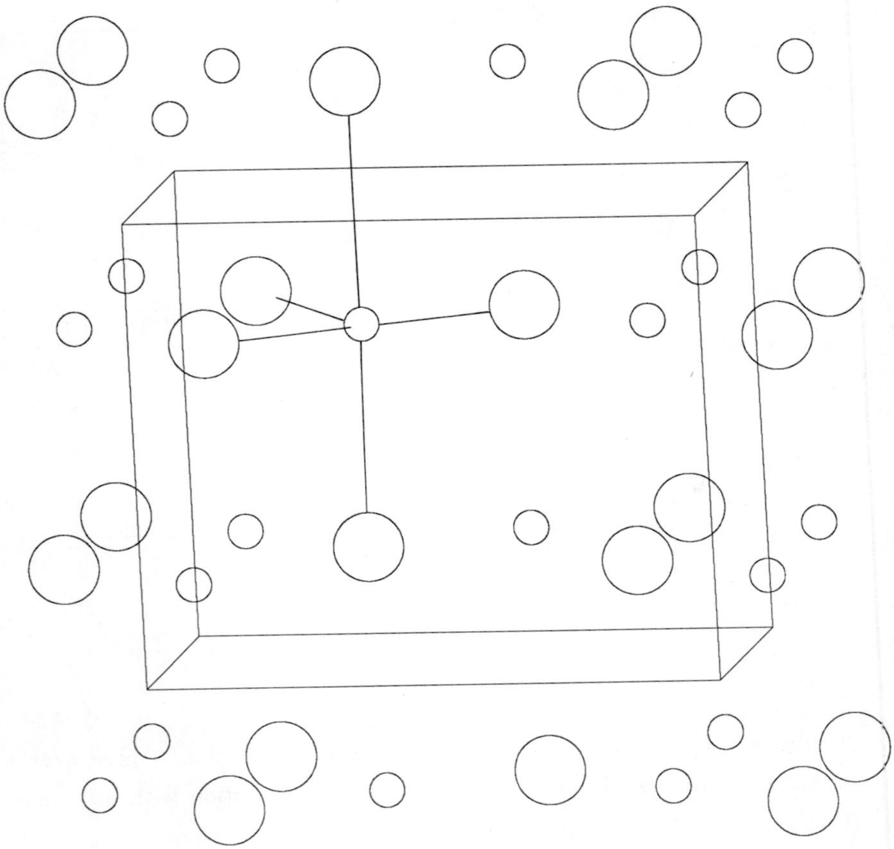

Figure 1.4b. Montroydite, HgO, perspective c-axis projection, a horizontal, b vertical. Each Hg (small circle) is in trigonal bipyramidal five-coordination with point symmetry m.

Table 1.4.2. Tenorite and Montroydite Cation Sites.

End-Member	Tenorite	Montroydite
C.N.	4	5
Occupant	Cu	Hg
Point Sym.	$\bar{1}$	m
Wyckoff Not.	4c	4c
Frac.Coord.		
x	1/4	.1150
y	1/4	1/4
z	0	.2450
Distances		
O1	(2)1.961	2.004
O2	(2)1.951	2.038
O3		2.847
O4		(2)2.825
Mean	1.956	2.508
σ	.006	.445
Poly.Vol.	planar	10.709
Site Energy	-1114.	-971.

Table 1.5.1. Corundum Group Unit Cells.

End-Member	Corundum	Hematite	Eskolaite	Karelianite
Formula	Al_2O_3	Fe_2O_3	Cr_2O_3	V_2O_3
Form.Wt.	101.961	159.692	151.990	149.882
Density	3.986	5.255	5.224	5.021
Mol Volume	25.577	30.388	29.093	29.850
Z	6	6	6	6
Cryst.Sys.	Trigonal	Trigonal	Trigonal	Trigonal
Cryst.Class	$\bar{3}m$	$\bar{3}m$	$\bar{3}m$	$\bar{3}m$
Space Group	$R\bar{3}c$	$R\bar{3}c$	$R\bar{3}c$	$R\bar{3}c$
Cell Parameters				
a	4.7589	5.038	4.9607	4.952
c	12.9912	13.772	13.599	14.002
Vol.	254.80	302.72	289.82	297.36
Ref.	Newnham & deHaan(1962)	Blake et al. (1966)	Newnham & deHaan(1962)	Newnham & deHaan(1962)

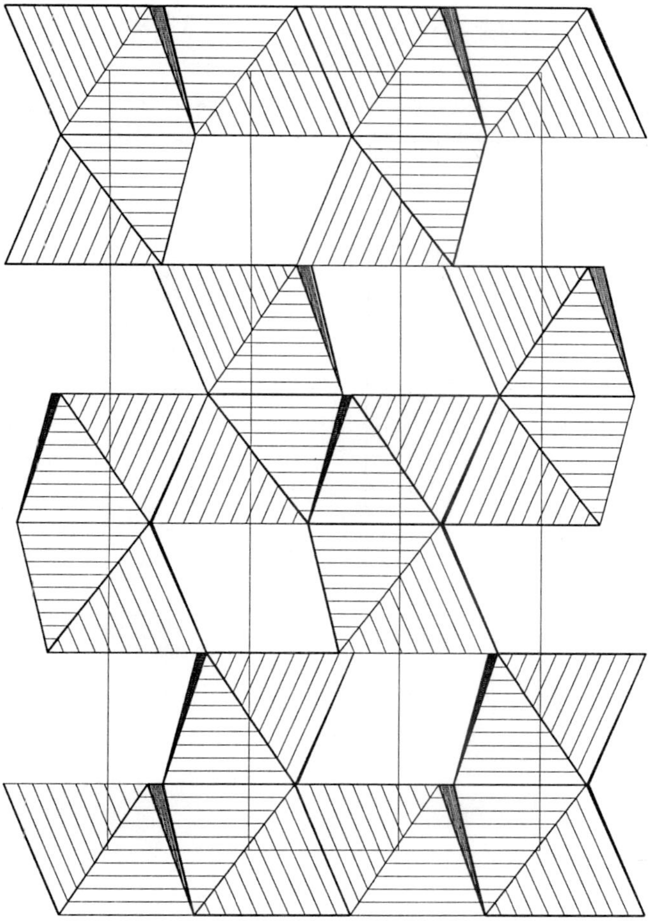

Figure 1.5. Corundum, Al_2O_3, [110] projection, c vertical. Each Al is in octahedral coordination with point symmetry 3. Each octahedron shares one face that is normal to c. This is the structure of several trivalent metal sesquioxides of which hematite (Fe_2O_3), eskolaite (Cr_2O_3), and karelianite (V_2O_3) occur as minerals. The structure of ilmenite ($FeTiO_3$) is a closely related ordered variant.

Table 1.5.2. Corundum Group Cation Sites.

End-Member	Corundum	Hematite	Eskolaite	Karelianite
C.N.	6	6	6	6
Occupant	Al	Fe	Cr	V
Point Sym.	3	3	3	3
Wyckoff Not.	12c	12c	12c	12c
Frac.Coord.				
x	0	0	0	0
y	0	0	0	0
z	.3520	.3553	.3475	.3463
Distances				
O1(3)	1.969	2.115	2.016	2.062
O1(3)	1.856	1.945	1.965	1.963
Mean	1.913	2.030	1.990	2.012
σ	.062	.093	.028	.054
Poly.Vol.	9.066	10.754	10.312	10.719
Q.E.	1.0200	1.0264	1.0131	1.0098
Ang. Var.	66.6	85.0	45.2	32.7
Site Energy	-2529.	-2401.	-2416.	-2309.

Table 1.6.1. Bixbyite Group Unit Cells.

End-Member	Bixbyite	Avicennite
Formula	$(Mn_{.983}Fe_{.017})_2O_3$	Tl_2O_3
Form.Wt.	157.905	456.738
Density	5.027	10.353
Mol Volume	31.412	44.115
Z	16	16
Cryst.Sys.	Isometric	Isometric
Cryst.Class	$m3$	$m3$
Space Group	$Ia3$	$Ia3$
Cell Parameters		
a	9.4146	10.543
Vol.	834.46	1171.91
Ref.	Geller (1971)	Papamantellos (1968)

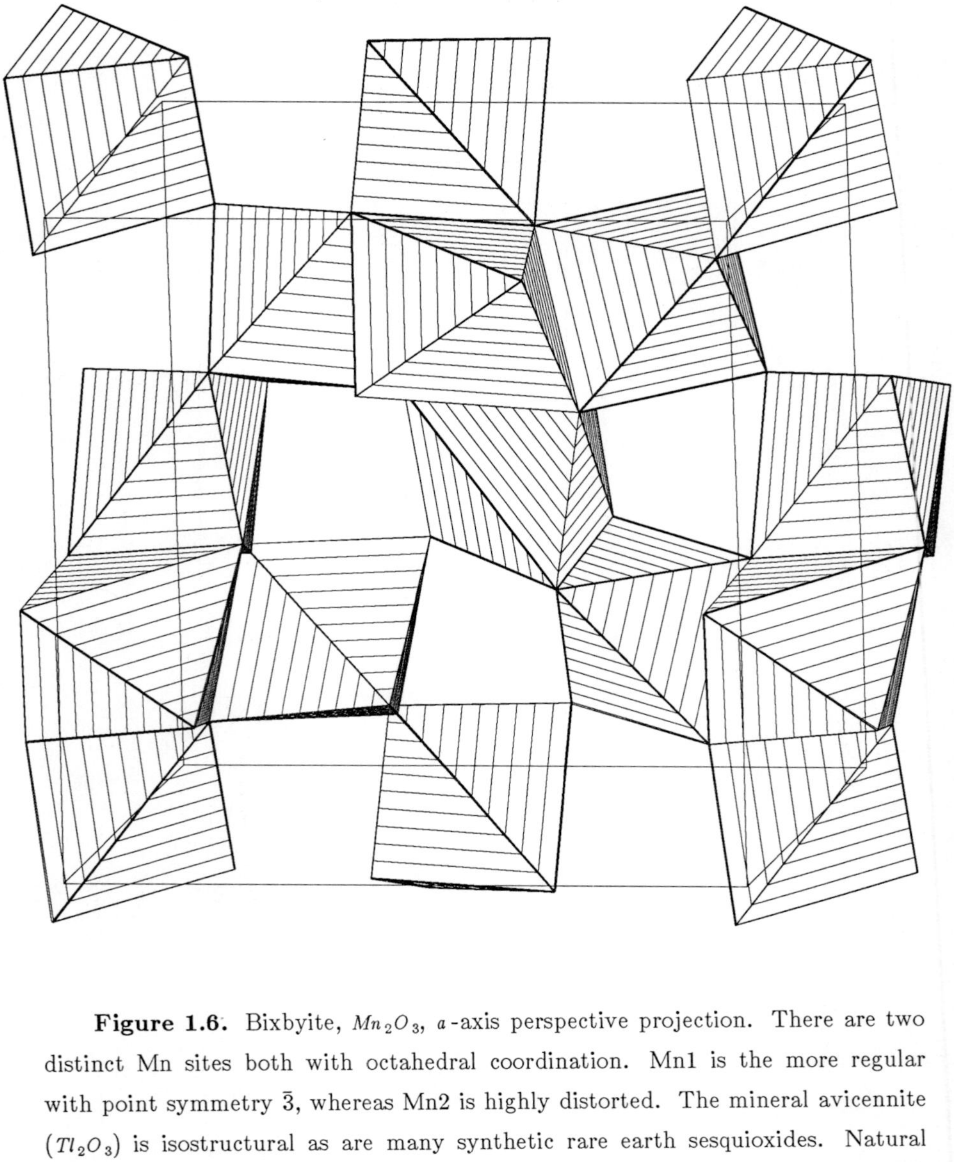

Figure 1.6. Bixbyite, Mn_2O_3, a-axis perspective projection. There are two distinct Mn sites both with octahedral coordination. Mn1 is the more regular with point symmetry $\bar{3}$, whereas Mn2 is highly distorted. The mineral avicennite (Tl_2O_3) is isostructural as are many synthetic rare earth sesquioxides. Natural bixbyite always contains some Fe, and pure Mn_2O_3 was found to have a subgroup symmetry, $Pcab$ (Geller, 1971).

Table 1.6.2. Bixbyite Group Cation Sites.

End-Member	Bixbyite		Avicennite	
Site	M1	M2	M1	M2
C.N.	6	6	6	6
Occupant	$Mn_{.983}Fe_{.017}$	$Mn_{.983}Fe_{.017}$	Tl	Tl
Point Sym.	$\bar{3}$	2	$\bar{3}$	2
Wyckoff Not.	8b	24d	8b	24d
Frac.Coord.				
x	1/4	-.030	1/4	-.029
y	1/4	0	1/4	0
z	1/4	1/4	1/4	1/4
Distances				
O1	(6)2.003	(2)1.927	(6)2.271	(2)2.140
O1		(2)2.084		(2)2.189
O1		(2)2.178		(2)2.475
Mean	2.003	2.063	2.271	2.268
σ	.000	.114	.000	.162
Poly.Vol.	10.580	10.759	15.046	13.764
Q.E.	1.0087	1.0605	1.0249	1.0895
Ang.Var.	28.9	181.8	77.8	224.3
Site Energy	-2468.	-2349.	-2160.	-2173.

Table 1.7.1. Arsenic and Antimony Sesquioxide Unit Cells.

End-Member	Arsenolite	Senarmontite	Claudetite	Valentinite
Formula	As_2O_3	Sb_2O_3	As_2O_3	Sb_2O_3
Form.Wt.	197.841	291.498	197.841	291.498
Density	3.870	5.583	3.960	5.844
Mol Volume	51.127	52.208	49.961	49.881
Z	16	16	4	4
Cryst.Sys.	Isometric	Isometric	Monoclinic	Orthorhombic
Cryst.Class	$m3m$	$m3m$	$2/m$	mmm
Space Group	$Fd3m$	$Fd3m$	$P2_1/n$	$Pccn$
Cell Parameters				
a	11.0744	11.1519	7.99	4.911
b	11.0744	11.1519	4.65	12.464
c	11.0744	11.1519	9.12	5.412
β			78.3	
Vol.	1358.19	1386.9	331.8	331.27
Ref.	Pertlik (1978)	Svensson (1975)	Pertlik (1975)	Svensson (1974)

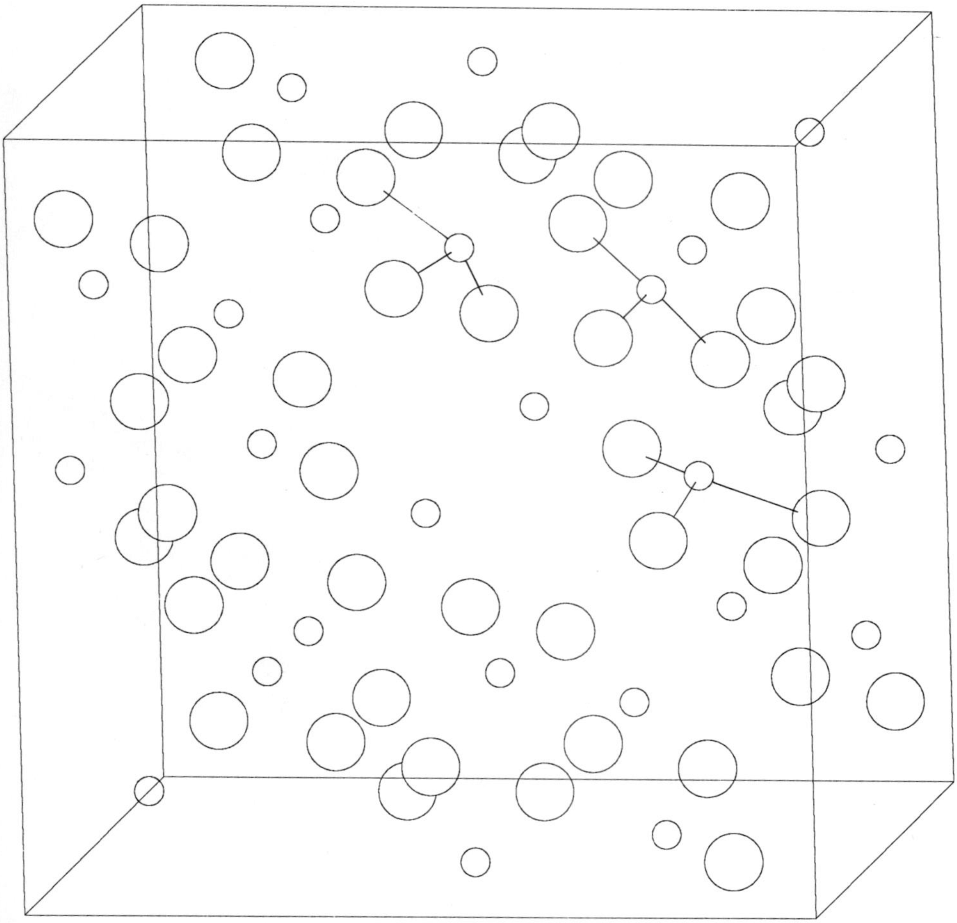

Figure 1.7a. Arsenolite, cubic As_2O_3, perspective a-axis projection. All As atoms are in identical three-coordination with point symmetry $3m$. Senarmontite, cubic Sb_2O_3, is isostructural.

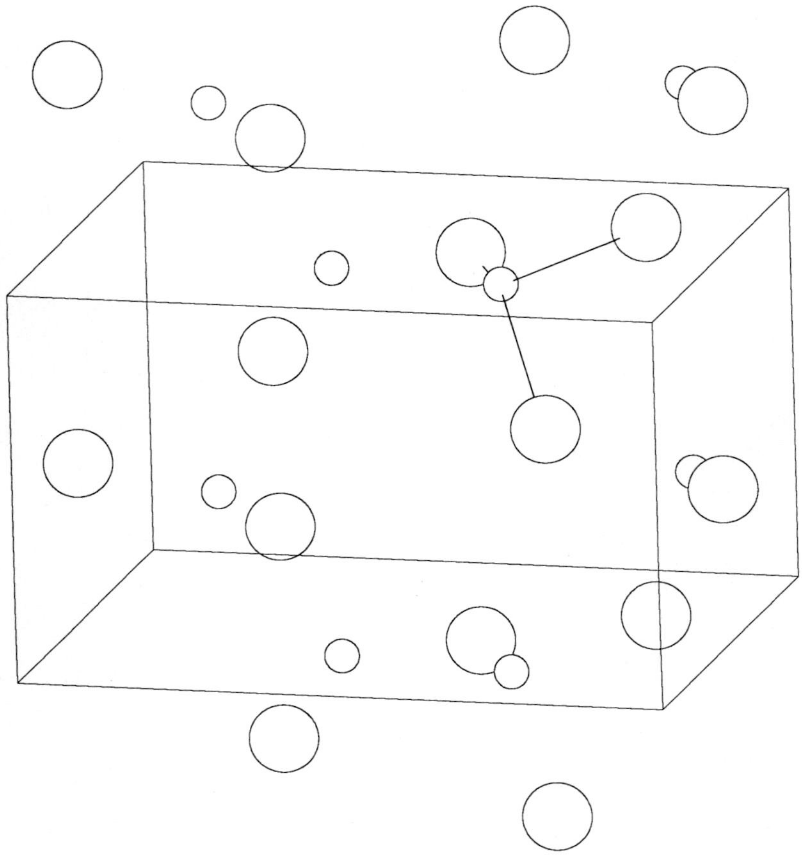

Figure 1.7b. Claudetite, monoclinic As_2O_3, perspective c-axis projection with a vertical and b horizontal. There are two distinct As atoms each in three-coordination.

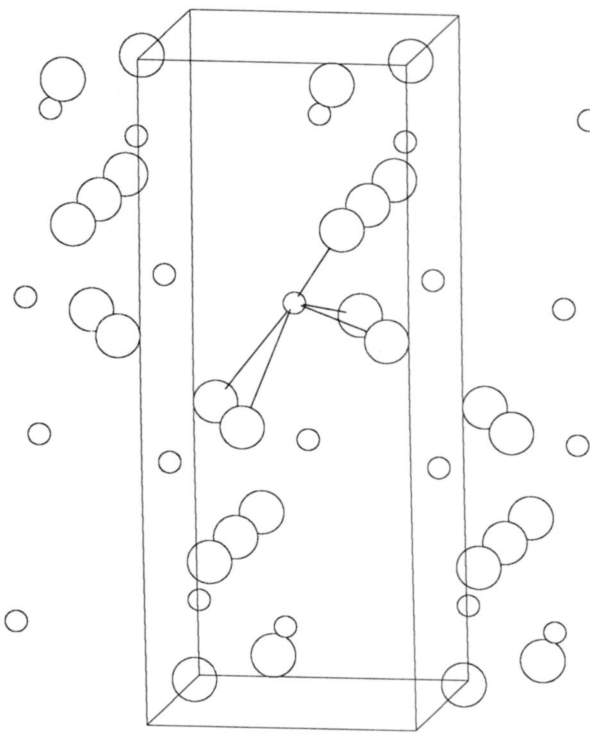

Figure 1.7c. Valentinite, orthorhombic Sb_2O_3, perspective c-axis projection, b vertical, a horizontal. Each Sb is identical in five-coordination.

Table 1.7.2. Arsenic and Antimony Sesquioxide Cation Sites.

End-Member	Arsenolite	Senarmontite	Claudetite		Valentinite
Site	As	Sb	As1	As2	Sb
C.N.	3	3	3	3	5
Occupant	As	Sb	As	As	Sb
Point Sym.	3m	3m	1	1	1
Wyckoff Not.	32e	32e	4e	4e	8e
Frac.Coord.					
x	.0221	.01027	.6163	.1841	.04149
y	.0221	.01027	.8311	.2910	.12745
z	.0221	.01027	.3013	.3717	.17845
Distances					
O1	(3)1.787	(3)1.9774	1.794	1.821	2.022
O2			1.796	1.771	2.619
O3			1.790	1.772	2.019
O4					2.519
O5					1.977
Mean	1.787	1.974	1.794	1.788	2.231
σ	.000	.000	.003	.029	.311
Poly.Vol.	-	-	-	-	5.468
Site Energy	-2419.	-2217.	-2267.	-2356.	-2184.

Table 1.8.1. Rutile Group Unit Cells.

End-Member	Rutile	Pyrolusite	Cassiterite	Stishovite
Formula	TiO_2	MnO_2	SnO_2	SiO_2
Form.Wt.	79.899	86.937	150.689	60.085
Density	4.2743	5.203	7.001	4.287
Mol Volume	18.693	16.708	21.523	14.017
Z	2	2	2	2
Cryst.Sys.	Tetragonal	Tetragonal	Tetragonal	Tetragonal
Cryst.Class	$4/mmm$	$4/mmm$	$4/mmm$	$4/mmm$
Space Group	$P\,4_2/mnm$	$P\,4_2/mnm$	$P\,4_2/mnm$	$P\,4_2/mnm$
Cell Parameters				
a	4.5845	4.396	4.737	4.1790
c	2.9533	2.871	3.185	2.6651
Vol.	62.07	55.48	71.47	46.54
Ref.	Shintani et al. (1975)	Kondrasev & Zaslavskij(1951)	Baur (1956)	Baur & Khan (1971)

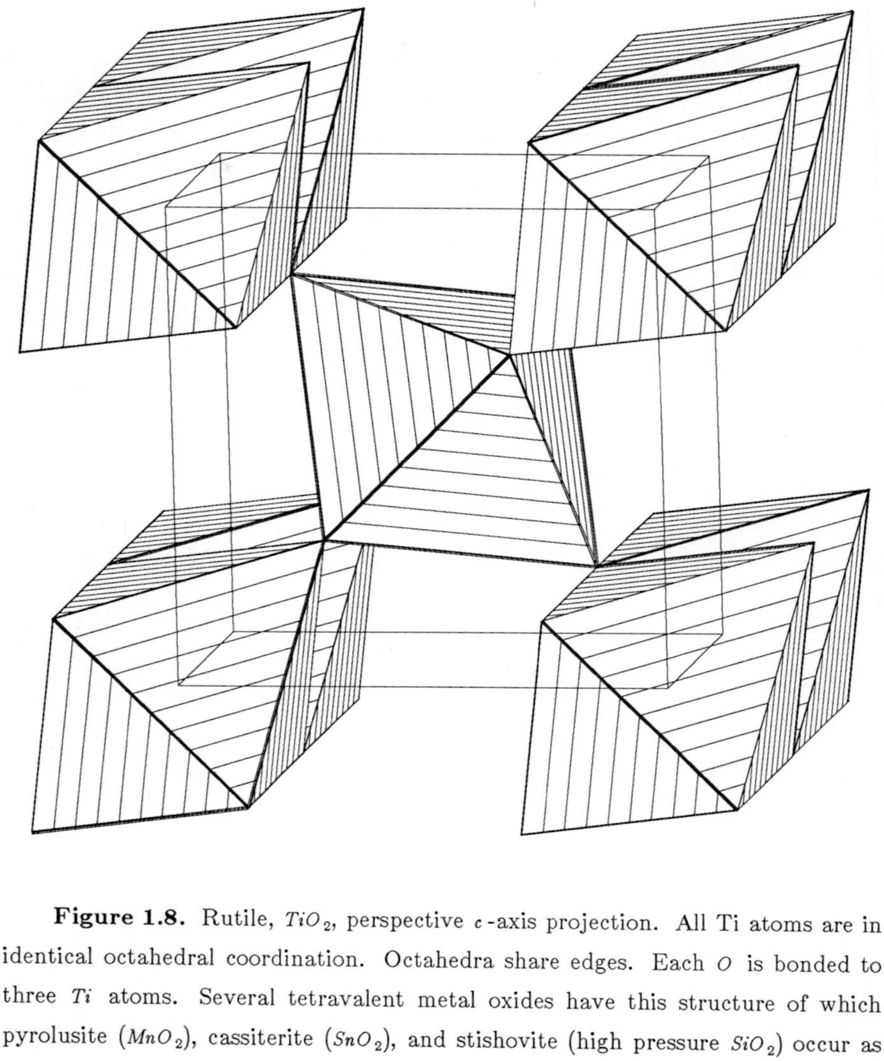

Figure 1.8. Rutile, TiO_2, perspective c-axis projection. All Ti atoms are in identical octahedral coordination. Octahedra share edges. Each O is bonded to three Ti atoms. Several tetravalent metal oxides have this structure of which pyrolusite (MnO_2), cassiterite (SnO_2), and stishovite (high pressure SiO_2) occur as minerals.

Table 1.8.2. Rutile Group Octahedral Sites.

End-Member	Rutile	Pyrolusite	Cassiterite	Stishovite
C.N.	6	6	6	6
Occupant	Ti	Mn^{+4}	Sn	Si
Point Sym.	*mmm*	*mmm*	*mmm*	*mmm*
Wyckoff Not.	2a	2a	2a	2a
Frac.Coord.				
x	0	0	0	0
y	0	0	0	0
z	0	0	0	0
Distances Cation-Ox				
1(2)	1.977	1.878	2.057	1.810
2(4)	1.944	1.891	2.051	1.757
Mean	1.955	1.887	2.053	1.775
σ	.017	.007	.003	.027
Poly.Vol.	9.846	8.847	11.292	7.365
Q.E.	1.0081	1.0079	1.0145	1.0080
Ang. Var.	28.4	28.0	51.1	27.1
Site Energy	-4133.	-4289.	-3953.	-4550.

Table 1.9.1. TiO_2 **Polymorph and Baddeleyite Unit Cells.**

Mineral	Rutile	Anatase	Brookite	Baddeleyite
Formula	TiO_2	TiO_2	TiO_2	ZrO_2
Form.Wt.	79.899	79.899	79.899	123.219
Density	4.2743	3.895	4.123	5.826
Mol Volume	18.693	20.516	19.377	21.149
Z	2	4	8	4
Cryst.Sys.	Tetragonal	Tetragonal	Orthorhombic	Monoclinic
Cryst.Class	$4/mmm$	$4/mmm$	mmm	$2/m$
Space Group	$P4_2/mnm$	$I4_1/amd$	$Pbca$	$P2_1/c$
Cell Parameters				
a	4.5845	3.7842	9.184	5.1454
b	4.5845	3.7842	5.447	5.2075
c	2.9533	9.5146	5.145	5.3107
β				99.23
Vol.	62.07	136.25	257.38	140.45
Ref.	Shintani et al. (1975)	Horn et al. (1972)	Baur (1961)	Smith & Newkirk (1965)

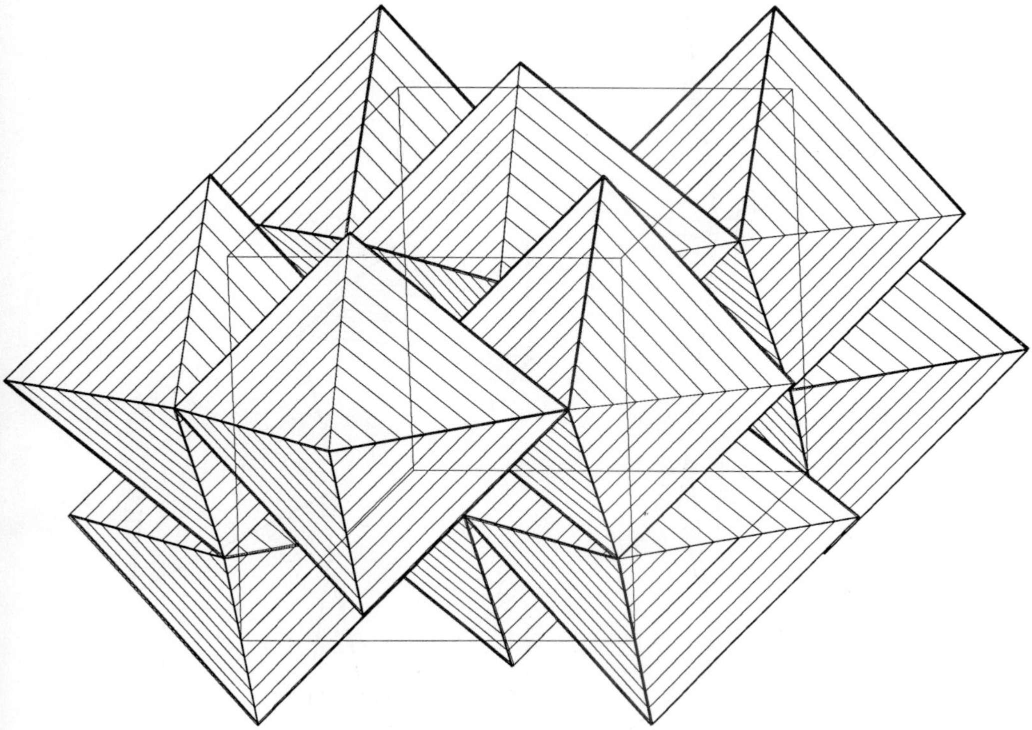

Figure 1.9a. Anatase, TiO_2, perspective c-axis projection. All Ti atoms are in identical octahedral coordination with point symmetry $\bar{4}2m$. Octahedra share edges.

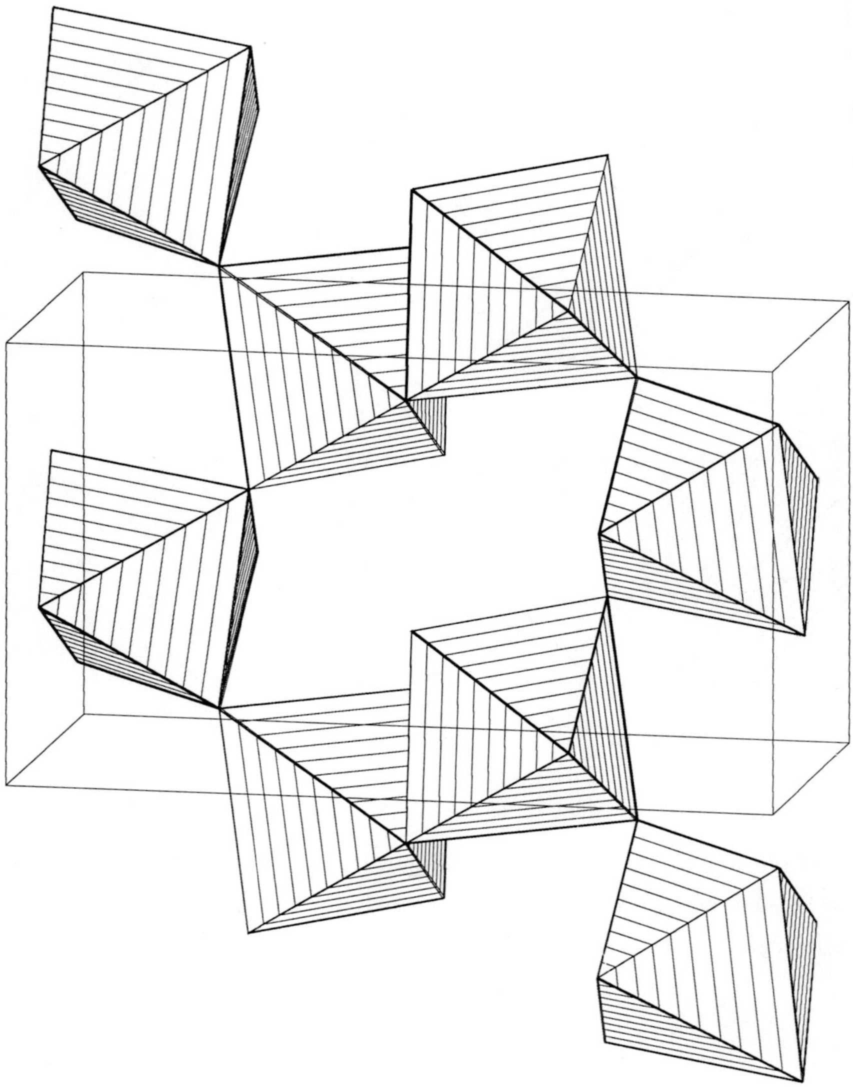

Figure 1.9b. Brookite, TiO_2, perspective c-axis projection, a horizontal. All atoms are in the general position with point symmetry 1. Each octahedron shares one edge.

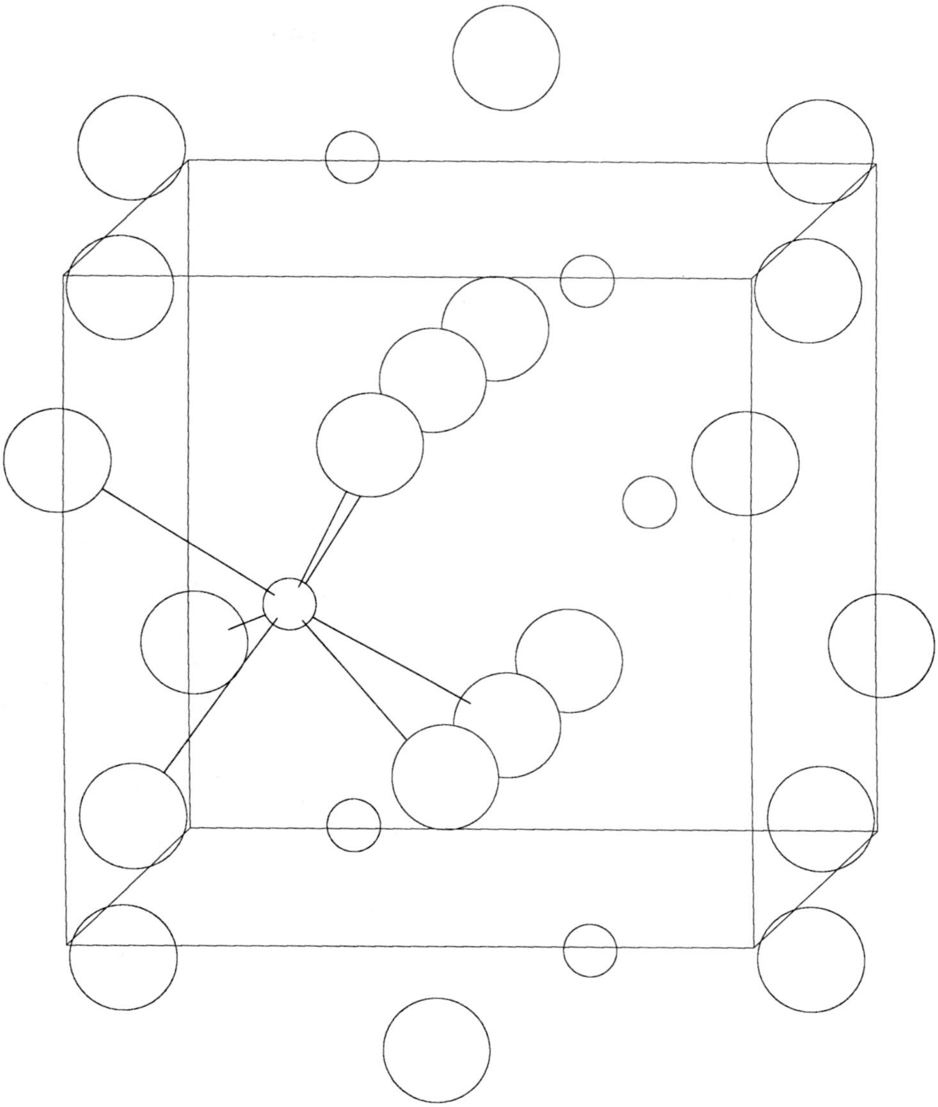

Figure 1.9c. Baddeleyite, ZrO_2, perspective c-axis projection, b horizontal. Each Zr atom is in irregular seven-coordination with point symmetry 1.

Table 1.9.2. TiO_2 **Polymorph and Baddeleyite Cation Sites.**

Mineral	Rutile	Anatase	Brookite	Baddeleyite
C.N.	6	6	6	7
Occupant	Ti	Ti	Ti	Zr
Point Sym.	*mmm*	$\bar{4}2m$	1	1
Wyckoff Not.	2a	4a	8c	4e
Frac.Coord.				
x	0	0	.1290	.2758
y	0	3/4	.0972	.0411
z	0	1/8	-.1371	.2082
Distances				
O1	(2)1.977	(2)1.964	1.993	2.051
O1	(4)1.944	(4)1.937	1.865	2.163
O1			1.993	2.057
O2			1.919	2.151
O2			2.046	2.285
O2			1.945	2.220
O2				2.189
Mean	1.955	1.946	1.959	2.159
sig.	.017	.014	.062	.084
Poly.Vol.	9.846	9.374	9.741	14.533
Q. E.	1.0081	1.0319	1.0204	--
Ang. Var.	28.4	113.7	68.6	--
Site Energy	-4133.	-4094.	-4107.	-3893.

Table 1.10.1. MnO_2 **Polymorph Unit Cells.**

End-Member	Pyrolusite	Ramsdellite
Formula	MnO_2	MnO_2
Form.Wt.	86.937	86.937
Density	5.203	4.874
Mol Volume	16.708	17.838
Z	2	4
Cryst.Sys.	Tetragonal	Orthorhombic
Cryst.Class	$4/mmm$	mmm
Space Group	$P\,4_2/mnm$	$Pnam$
Cell Parameters		
a	4.396	9.32
b	4.396	4.46
c	2.871	2.850
Vol.	55.48	118.47
Ref.	Kondrasev & Zaslavskij(1951)	Kondrasev & Zaslavskij(1951)

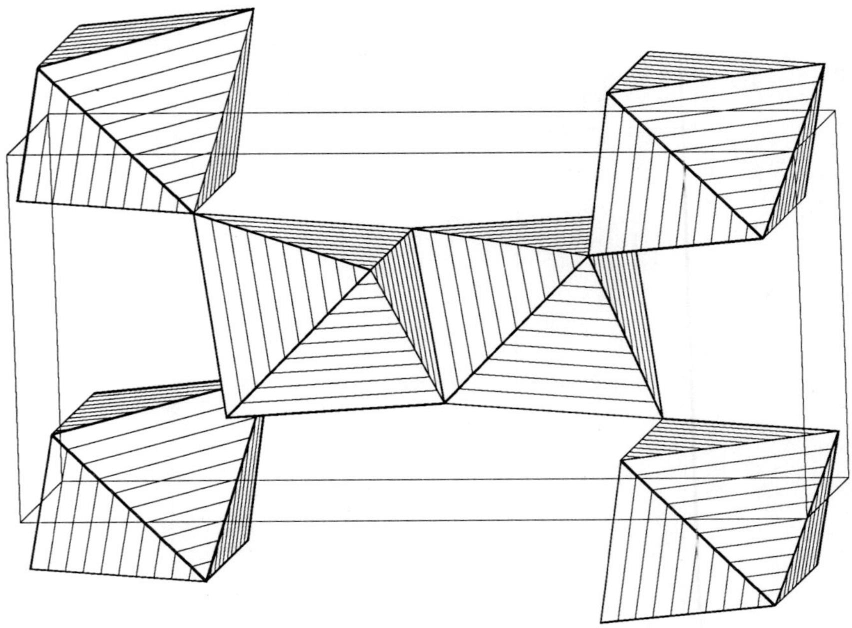

Figure 1.10. Ramsdellite, MnO_2, perspective c-axis projection. All Mn atoms are in identical octahedral coordination with point symmetry m.

Table 1.10.2. MnO_2 **Polymorph Mn Sites.**

End-Member	Pyrolusite	Ramsdellite
C.N.	6	6
Occupant	Mn	Mn
Point Sym.	*mmm*	*m*
Wyckoff Not.	2a	4c
Frac.Coord.		
x	0	.140
y	0	.020
z	0	1/4
Distances		
O1	(2)1.878	(2)1.949
O1	(4)1.891	(1)1.887
O2		(2)1.861
O2		(1)1.837
Mean	1.887	1.891
σ	.007	.048
Poly.Vol.	8.847	8.798
Q.E.	1.0079	1.0169
Ang.Var.	28.0	54.1
Site Energy	-4289.	-4127.

Table 1.11.1. Uraninite Group Unit Cells.

End-Member	Uraninite	Thorianite
Formula	UO_2	ThO_2
Form.Wt.	270.029	264.037
Density	10.968	9.987
Mol Volume	24.620	26.439
Z	4	4
Cryst.Sys.	Isometric	Isometric
Cryst.Class	$m3m$	$m3m$
Space Group	$Fm3m$	$Fm3m$
Cell Parameters		
a	5.4682	5.5997
Vol.	163.51	175.59
Ref.	Leonova (1959)	Vogel & Kempter (1959)

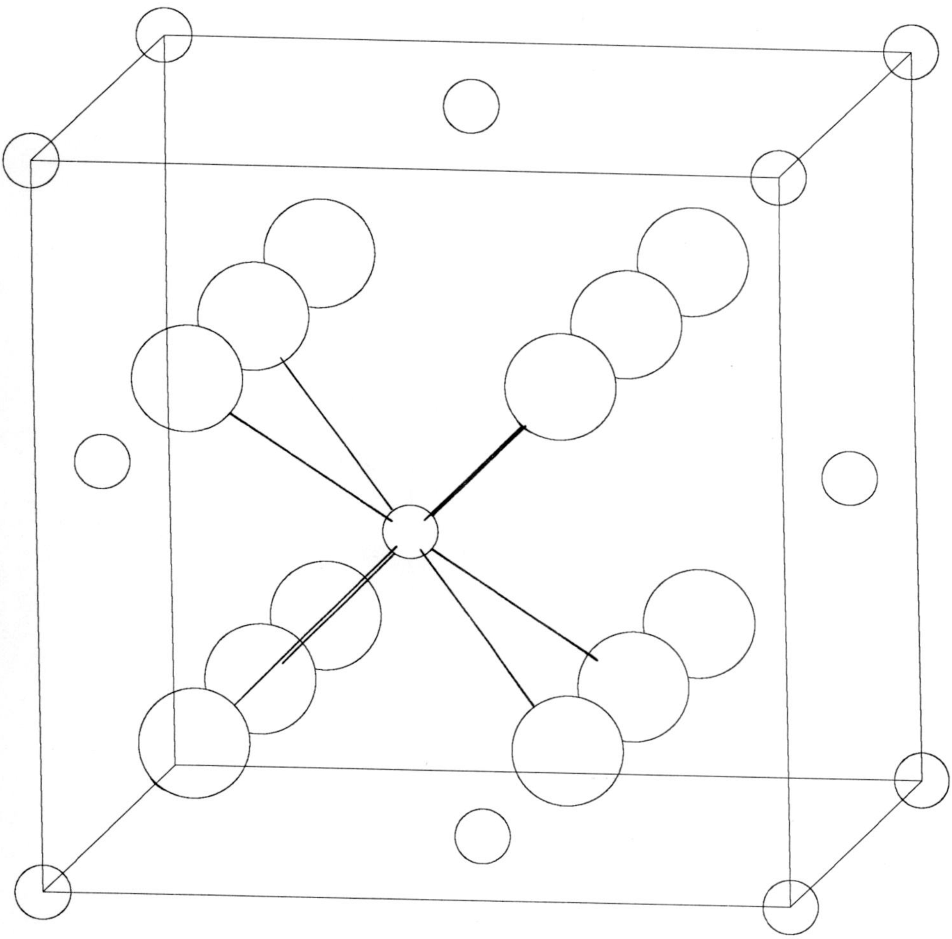

Figure 1.11. Uraninite, UO_2, perspective a-axis projection. Each U atom is in perfectly regular cubic eight coordination with point symmetry $m3m$. Each O is in regular tetrahedral coordination. Thorianite (ThO_2) and fluorite (CaF_2) are isostructural.

Table 1.11.2. Uraninite Group Cation Sites.

End-Member	Uraninite	Thorianite
C.N.	8	8
Occupant	U	Th
Point Sym.	$m3m$	$m3m$
Wyckoff Not.	4b	4b
Frac.Coord.		
x	0	0
y	0	0
z	0	0
Distances		
O1(8)	2.368	2.425
Poly.Vol.	40.876	43.897
Site Energy	-3396.	-3316.

Table 1.12.1. Tellurite and Paratellurite Unit Cells.

End-Member	Tellurite	Paratellurite
Formula	TeO_2	TeO_2
Form.Wt.	159.599	159.599
Density	5.749	6.043
Mol Volume	27.759	26.412
Z	8	4
Cryst.Sys.	Orthorhombic	Tetragonal
Cryst.Class	mmm	422
Space Group	$Pbca$	$P4_12_12$
Cell Parameters		
a	12.035	4.796
b	5.464	4.796
c	5.607	7.626
Vol.	368.71	175.41
Ref.	Beyer (1967)	Leciejewicz (1961)

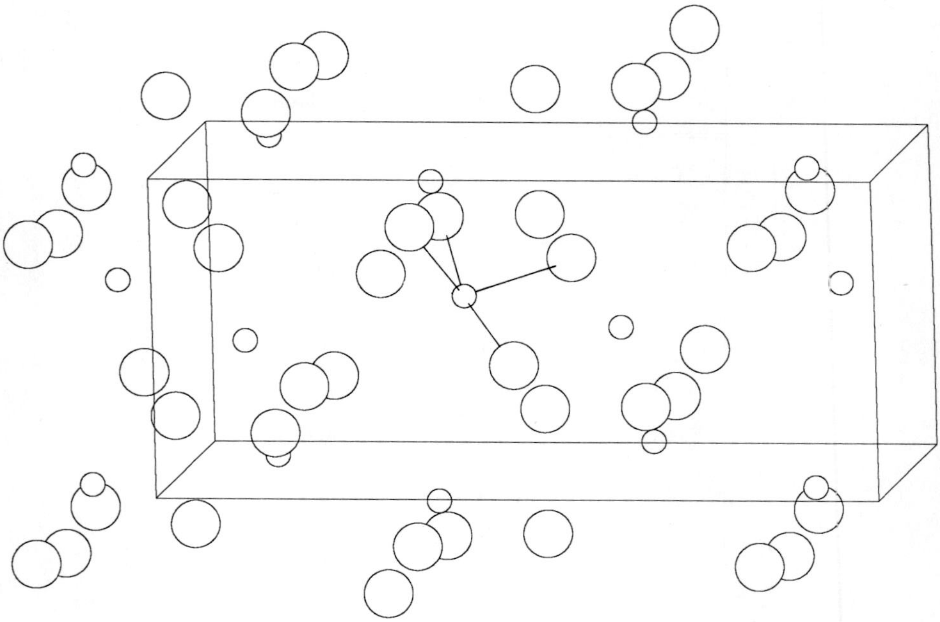

Figure 1.12a. Tellurite, TeO_2, perspective c-axis projection, a horizontal. Each Te is in identical irregular four-coordination with point symmetry 1.

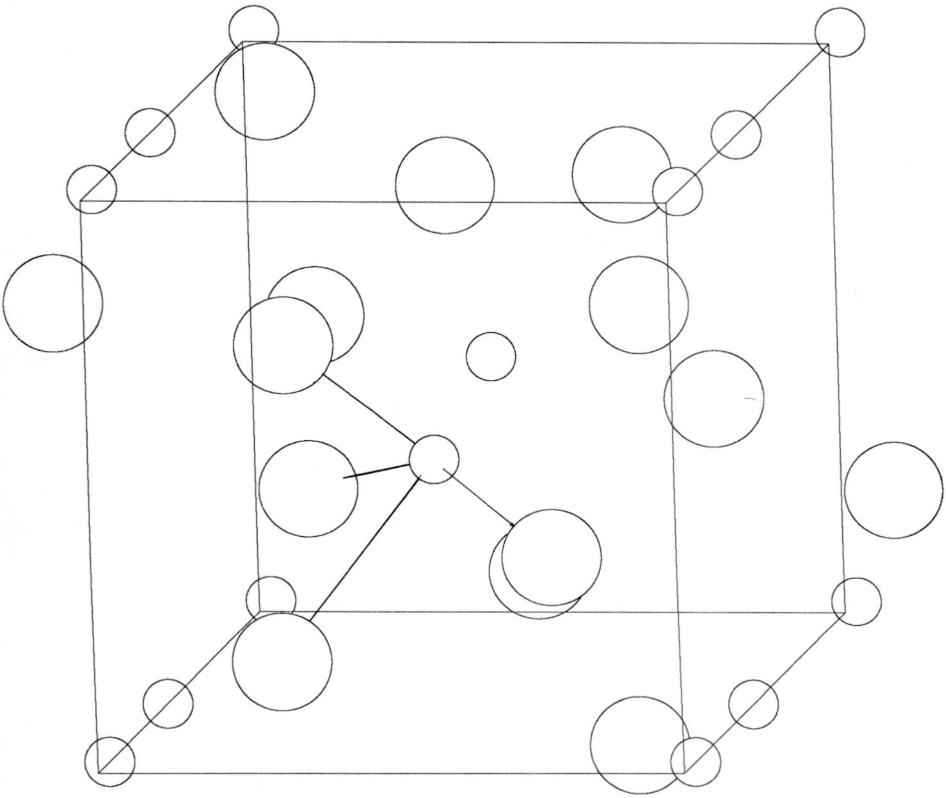

Figure 1.12b. Paratellurite, TeO_2, perspective c-axis projection. Each Te (small circle) is in identical, irregular four-coordination with point symmetry 2.

Table 1.12.2. Tellurite and Paratellurite Cation Sites.

End-Member	Tellurite	Paratellurite
C.N.	4	4
Occupant	Te	Te
Point Sym.	1	2
Wyckoff Not.	8d	4a
Frac.Coord.		
x	.1182	.0200
y	.0255	.0200
z	.3781	0
Distances		
O1	1.877	(2)1.919
O2	2.196	(2)2.087
O3	1.927	
O4	2.070	
Mean	2.018	2.003
σ	.144	.097
Poly.Vol.	2.494	2.508
Q.E.	1.4240	1.3950
Ang.Var.	544.	651.
Site Energy	-3758.	-3784.

Chapter 2

DOUBLE OXIDES

Table 2.1.1. Ilmenite Group Unit Cells.

End-Member	Ilmenite	Pyrophanite	$MgSiO_3$
Formula	$FeTiO_3$	$MnTiO_3$	$MgSiO_3$
Form.Wt.	151.745	150.836	100.396
Density	4.786	4.603	3.810
Mol Volume	31.705	32.766	26.354
Z	6	6	6
Cryst.Sys.	Trigonal	Trigonal	Trigonal
Cryst.Class	$\bar{3}$	$\bar{3}$	$\bar{3}$
Space Group	$R\bar{3}$	$R\bar{3}$	$R\bar{3}$
Cell Parameters			
a	5.0884	5.137	4.7284
c	14.0855	14.283	13.5591
Vol.	315.84	326.41	262.54
Ref.	Wechsler & Prewitt(1984)	Wyckoff (1963)	Horiuchi et al. (1982)

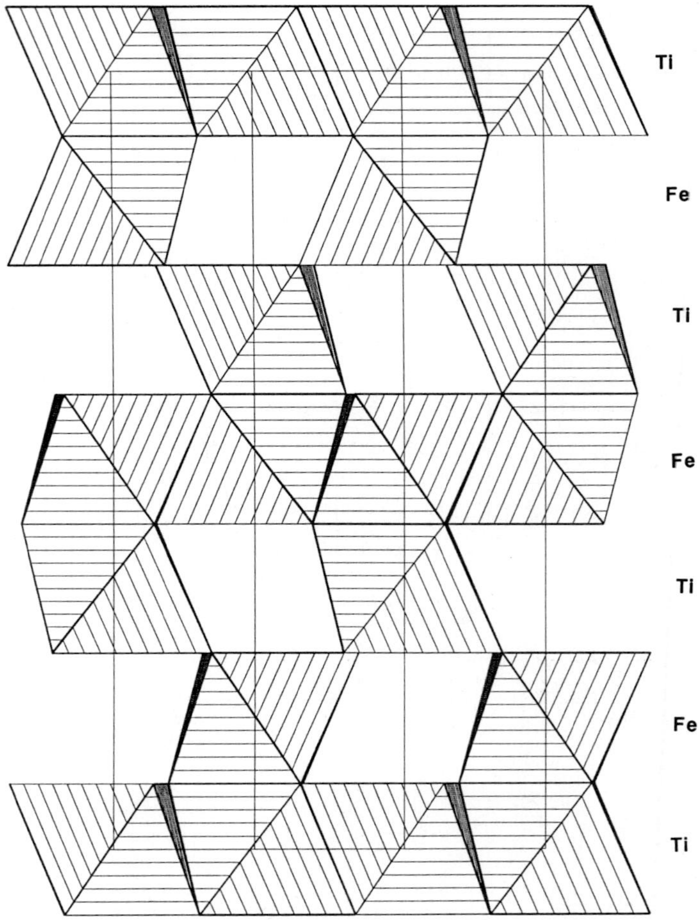

Figure 2.1. Ilmenite, $FeTiO_3$, [110] projection, c vertical. Layers of Fe and Ti octahedra alternate along c. This structure is an ordered variant of corundum (Fig. 1.5.).

Table 2.1.2. Ilmenite Group Divalent Sites.

End-Member	Ilmenite	Pyrophanite	$MgSiO_3$
C.N.	6	6	6
Occupant	Fe	Mn	Mg
Point Sym.	3	3	3
Wyckoff Not.	6c	6c	6c
Frac.Coord.			
x	0	0	0
y	0	0	0
z	.35537	.3570	.3597
Distances			
1(3)	2.201	2.230	2.163
2(3)	2.078	2.024	1.990
Mean	2.081	2.127	2.076
σ	.068	.113	.095
Poly.Vol.	12.562	12.336	11.238
Q.E.	1.0271	1.0289	1.0429
Ang. Var.	91.8	91.8	143.4
Site Energy	-1179.	-1220.	-1183.

Table 2.1.3. Ilmenite Group Tetravalent Sites.

End-Member	Ilmenite	Pyrophanite	$MgSiO_3$
C.N.	6	6	6
Occupant	Ti	Ti	Si
Point Sym.	3	3	3
Wyckoff Not.	6c	6c	6c
Frac.Coord.			
x	0	0	0
y	0	0	0
z	.14640	.1430	.15768
Distances			
1(3)	2.089	2.190	1.830
2(3)	1.874	1.912	1.768
Mean	1.982	2.051	1.799
sig.	.117	.152	.034
Poly.Vol.	10.001	11.067	7.592
Q.E.	1.0277	1.0310	1.0152
Ang. Var.	86.0	91.4	52.8
Site Energy	-3959.	-3809.	-4352.

Table 2.2.1. Perovskite Group Unit Cells.

End-Member	Perovskite	$MgSiO_3$
Formula	$CaTiO_3$	$MgSiO_3$
Form.Wt.	135.98	100.396
Density	4.044	4.107
Mol Volume	33.63	24.445
Z	4	4
Cryst.Sys.	Orthorhombic	Orthorhombic
Cryst.Class	*mmm*	*mmm*
Space Group	*Pbnm*	*Pbnm*
Cell Parameters		
a	5.3670	4.7754
b	5.4439	4.9292
c	7.6438	6.8969
Vol.	223.33	162.35
Ref.	Kay & Bailey(1957)	Horiuchi et al. (1987)

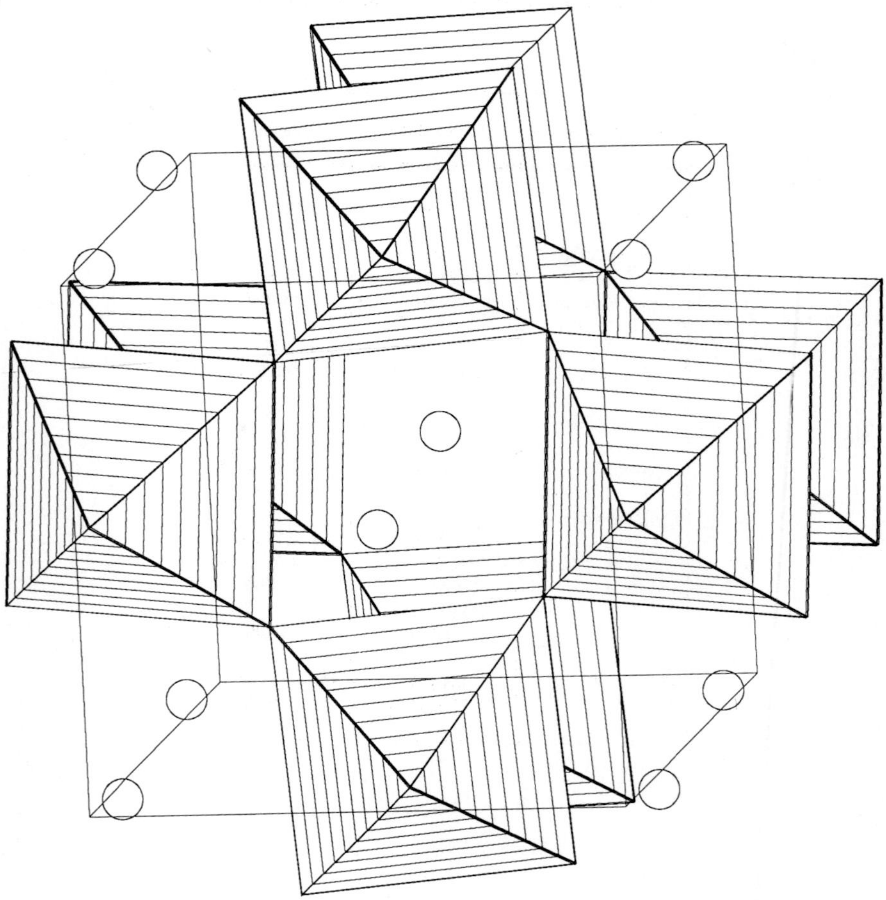

Figure 2.2. Perovskite, $CaTiO_3$, perspective c-axis projection, a horizontal. Ca is in irregular ten-coordination with point symmetry m. Ti is in octahedral coordination with point symmetry $\bar{1}$. In the $MgSiO_3$ polymorph with this structure, Mg is in the Ca site, but the coordination number is reduced to eight by rotation of the octahedra.

Table 2.2.2. Perovskite Group Cation Sites.

End-Member	Perovskite		$MgSiO_3$	
C.N.	10	6	8	6
Occupant	Ca	Ti	Mg	Si
Point Sym.	m	$\bar{1}$	m	$\bar{1}$
Wyckoff Not.	4c	4b	4c	4b
Frac.Coord.				
x	.000	0	.974	0
y	.030	1/2	.063	1/2
z	1/4	0	1/4	0
Distances Cation-Ox				
O1	(1)2.794	(2)1.924	(1)2.014	(2)1.801
O1	(1)2.664		(1)2.097	
O1	(1)2.883			
O1	(1)2.486			
O2	(2)2.584	(2)1.924	(2)2.278	(2)1.782
O2	(2)2.553	(2)1.928	(2)2.052	(2)1.796
O2	(2)2.685		(2)2.427	
Mean	2.647	1.926	2.203	1.793
σ	.121	.002	.171	.008
Poly.Vol.	37.112	9.492	20.100	7.681
Q.E.		1.0019		1.0005
Ang.Var.		6.7		1.6
Site Energy	-950.	-4256.	-1143.	-4518.

Table 2.3.1. Oxide Spinel Group Unit Cells.

End-Member	Spinel	Hercynite	Magnesio-ferrite	Magnetite	Jacobsite	Magnesio-chromite	Chromite	Ulvospinel
Formula	$MgAl_2O_4$	$FeAl_2O_4$	$MgFe_2O_4$	$FeFe_2O_4$	$MnFe_2O_4$	$MgCr_2O_4$	$FeCr_2O_4$	$TiFe_2O_4$
Form.Wt.	142.273	173.808	200.004	231.539	230.630	192.302	223.837	223.592
Density	3.578	4.256	4.547	5.200	4.969	4.414	5.054	4.775
Mol Volume	39.762	40.843	43.989	44.528	46.416	43.564	44.293	46.826
X_{inv}	0.07	0.00	0.90	1.00	0.15	0.00	0.00	1.00
Z	8	8	8	8	8	8	8	8
Cryst.Sys.	Isometric	Isometric	Isometric	Isometric	Isometric	Isometric	Isometric	Isometric
Cryst.Class	$m3m$	$m3m$	$m3m$	$m3m$	$m3m$	$m3m$	$m3m$	$m3m$
Space Group	$Fd3m$	$Fd3m$	$Fd3m$	$Fd3m$	$Fd3m$	$Fd3m$	$Fd3m$	$Fd3m$
Cell Parameters								
a	8.0832	8.1558	8.360	8.394	8.5110	8.333	8.3792	8.536
Vol.	528.14	542.50	584.28	591.43	616.51	578.63	588.31	621.96
Ref.	Fischer (1967)	Hill (1984)	Hill et al. (1979)	Hill et al. (1979)	Hill et al. (1979)	Hill et al. (1979)	Hill et al. (1979)	Ishikawa et al.(1972)

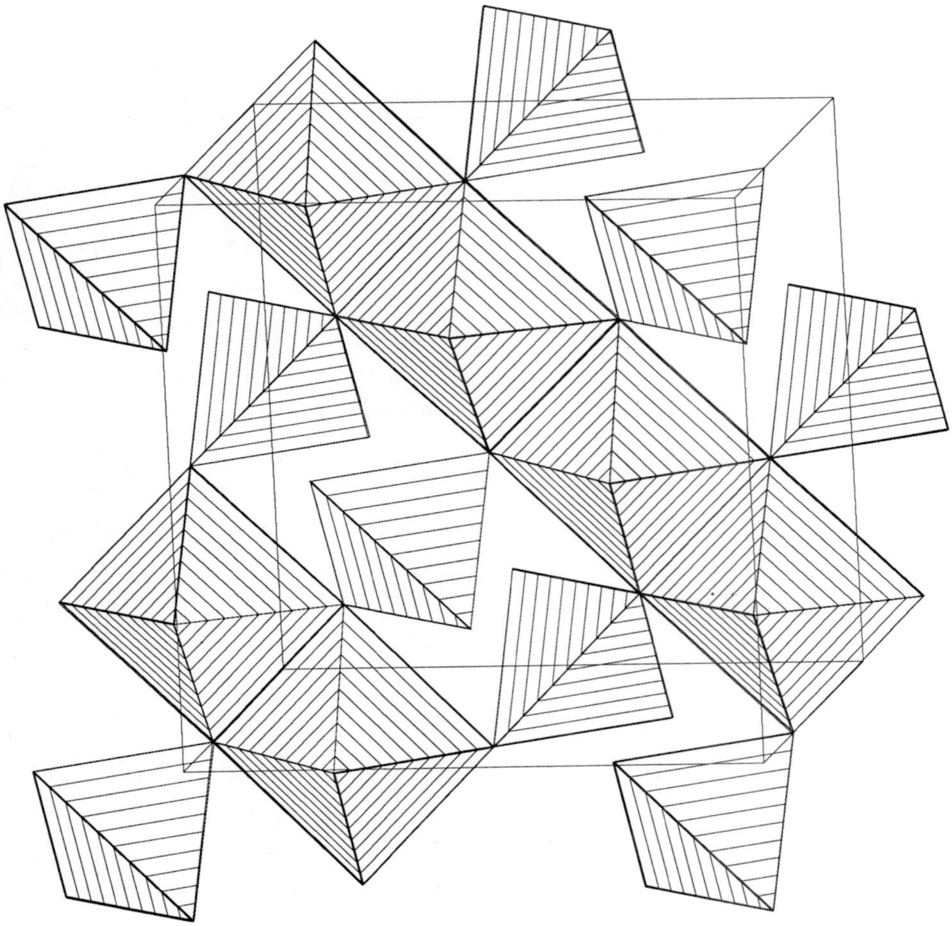

Figure 2.3. Spinel, $MgAl_2O_4$, perspective a-axis projection. Oxygen is in an approximately close-packed arrangement, and there are half as many tetrahedral as octahedral sites. The octahedron has point symmetry $\bar{3}m$ and the tetrahedron $\bar{4}3m$.

Table 2.3.2. Oxide Spinel Octahedral Sites.

End-Member	Spinel	Hercynite	Magnesio-ferrite	Magnetite	Jacobsite	Magnesio-chromite	Chromite	Ulvospinel
C.N.	6	6	6	6	6	6	6	6
Occupant	$Al_{.96}Mg_{.04}$	Al	$Mg_{.45}Fe^{+3}_{.55}$	$Fe^{+2}_{.5}Fe^{+3}_{.5}$	$Fe^{+3}_{.93}Mn_{.07}$	Cr^{+3}	Cr^{+3}	$Fe^{+2}_{.5}Ti_{.5}$
Point Sym.	$\bar{3}m$	$\bar{3}m$	$\bar{3}m$	$\bar{3}m$	$\bar{3}m$	$\bar{3}m$	$\bar{3}m$	$\bar{3}m$
Wyckoff Not.	16d	16d	16d	16d	16d	16d	16d	16d
Frac.Coord.								
x	1/2	1/2	1/2	1/2	1/2	1/2	1/2	1/2
y	1/2	1/2	1/2	1/2	1/2	1/2	1/2	1/2
z	1/2	1/2	1/2	1/2	1/2	1/2	1/2	1/2
Distances								
O(6)	1.926	1.937	2.033	2.059	2.035	1.994	1.990	2.044
Poly.Vol.	9.371	9.505	11.151	11.612	11.074	10.440	10.322	11.252
Q.E.	1.0108	1.0125	1.0033	1.0015	1.0092	1.0087	1.0123	1.0084
Ang. Var.	40.8	47.4	12.2	5.6	34.7	32.8	46.7	31.6
Site Energy	-2407.	-2444.	-1791.	-1701.	-2226.	-2353.	-2377.	-2293.
Model Charge	2.96	3.0	2.55	2.5	2.925	3.0	3.0	3.0

Table 2.3.3. Oxide Spinel Tetrahedral Sites.

End-Member	Spinel	Hercynite	Magnesio-ferrite	Magnetite	Jacobsite	Magnesio-chromite	Chromite	Ulvospinel
C.N.	4	4	4	4	4	4	4	4
Occupant	$Mg_{.93}Al_{.07}$	Fe^{+2}	$Fe^{+3}_{.90}Mg_{.10}$	Fe^{+3}	$Mn^{+2}_{.85}Fe^{+3}_{.15}$	Mg	Fe^{+2}	Fe^{+2}
Point Sym.	$\bar{4}3m$	$\bar{4}3m$	$\bar{4}3m$	$\bar{4}3m$	$\bar{4}3m$	$\bar{4}3m$	$\bar{4}3m$	$\bar{4}3m$
Wyckoff Not.	8a	8a	8a	8a	8a	8a	8a	8a
Frac.Coord.								
x	1/8	1/8	1/8	1/8	1/8	1/8	1/8	1/8
y	1/8	1/8	1/8	1/8	1/8	1/8	1/8	1/8
z	1/8	1/8	1/8	1/8	1/8	1/8	1/8	1/8
Distances								
O(4)	1.924	1.954	1.911	1.887	2.012	1.966	2.006	2.011
Poly.Vol.	3.653	3.827	3.584	3.449	4.181	3.899	4.141	4.172
Q.E.	1.0000	1.0000	1.0000	1.0000	1.0000	1.0000	1.0000	1.0000
Ang.Var.	0.0	0.0	0.0	0.0	0.0	0.0	0.0	0.0
Site Energy	-1269.	-1179.	-2211.	-2384.	-1293.	-1189.	-1149.	-1164.
Model Charge	2.07	2.0	2.90	3.0	2.15	2.0	2.0	2.0

Table 2.4.1. Pseudobrookite Group Unit Cells.

End-Member	Pseudobrookite	Tialite	Armalcolite
Formula	Fe_2TiO_5	Al_2TiO_5	$(Mg_{.5}Fe_{.5})Ti_2O_5$
Form.Wt.	239.591	181.860	215.876
Density	4.406	3.702	3.904
Mol Volume	54.375	49.128	55.298
Z	4	4	4
Cryst.Sys.	Orthorhombic	Orthorhombic	Orthorhombic
Cryst.Class	*mmm*	*mmm*	*mmm*
Space Group	*Bbmm*	*Bbmm*	*Bbmm*
Cell Parameters			
a	9.767	9.429	9.7762
b	9.947	9.636	10.0214
c	3.717	3.591	3.7485
Vol.	361.12	326.27	367.25
Ref.	Pauling (1930) cell: Akimoto (1957)	Morosin & Lynch (1972)	Wechsler et al. (1976)

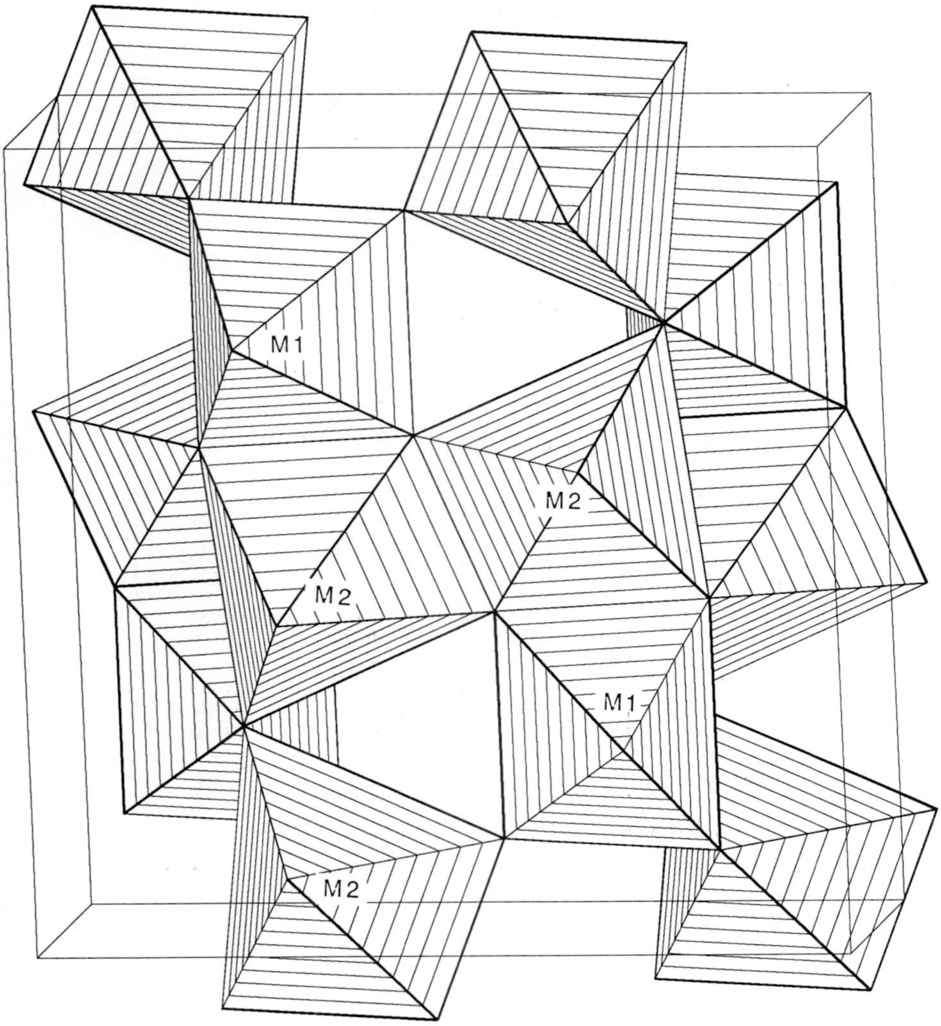

Figure 2.4. Pseudobrookite, Fe_2TiO_5, perspective c-axis projection with a horizontal. There are two distinct octahedral sites, M1 with point symmetry mm, and M2 with point symmetry m. In pseudobrookite and tialite (Al_2TiO_5) M2 is slightly larger, whereas in armalcolite ($(Mg,Fe)Ti_2O_5$), M1 is slightly larger. Ferropseudobrookite ($FeTi_2O_5$) is also isostructural, but no refinement of an end-member structure was available.

Table 2.4.2. Pseudobrookite Group M1 Sites.

End-Member	Pseudobrookite	Tialite	Armalcolite
C.N.	6	6	6
Occupant	Ti	Ti	$Fe_{.5}Mg_{.5}$
Point Sym.	*mm*	*mm*	*mm*
Wyckoff Not.	4c	4c	4c
Frac.Coord.			
x	.190	.1854	.19223
y	1/4	1/4	1/4
z	0	0	0
Distances			
O1(2)	1.899	1.921	2.036
O2(2)	1.986	1.821	1.965
O3(2)	1.937	2.092	2.193
Mean	1.941	1.944	2.065
σ	.039	.122	.104
Poly.Vol.	9.410	8.935	10.418
Q.E.	1.0239	1.0672	1.0848
Ang.Var.	76.1	181.8	230.85
Site Energy	-4093.	-4052.	-1248.
Model Charge	4.0	4.0	2.0

Table 2.4.3. Pseudobrookite Group M2 Sites.

End-Member	Pseudobrookite	Tialite	Armalcolite
C.N.	6	6	6
Occupant	Fe^{+3}	Al	Ti
Point Sym.	*m*	*m*	*m*
Wyckoff Not.	8f	8f	8f
Frac.Coord.			
x	.135	.13478	.13479
y	.560	.56150	.56447
z	0	0	0
Distances			
O1	2.304	2.080	2.064
O2	1.906	1.900	1.991
O2	1.827	1.808	1.845
O3	2.302	2.113	2.176
O3(2)	1.966	1.866	1.943
Mean	2.045	1.939	1.993
σ	.206	.126	.114
Poly.Vol.	10.468	9.188	10.0136
Q.E.	1.0678	1.0418	1.0391
Ang.Var.	207.0	128.1	121.5
Site Energy	-2395.	-2545.	-3941.
Model Charge	3.0	3.0	4.0

Table 2.5.1. Tungstate Unit Cells.

End-Member	Ferberite	Huebnerite	Scheelite
Formula	$FeWO_4$	$MnWO_4$	$CaWO_4$
Form.Wt.	303.695	302.786	287.928
Density	7.549	7.265	6.115
Mol Volume	40.228	41.676	47.087
Z	2	2	4
Cryst.Sys.	Monoclinic	Monoclinic	Tetragonal
Cryst.Class	$2/m$	$2/m$	$4/m$
Space Group	$P\,2/c$	$P\,2/c$	$I\,4_1/a$
Cell Parameters			
a	4.730	4.8238	5.243
b	5.703	5.7504	
c	4.952	4.9901	11.376
β	90.00	91.18	
Vol.	133.58	138.39	312.72
Ref.	Uelkue (1967)	Weitzel (1976)	Kay et al. (1964)

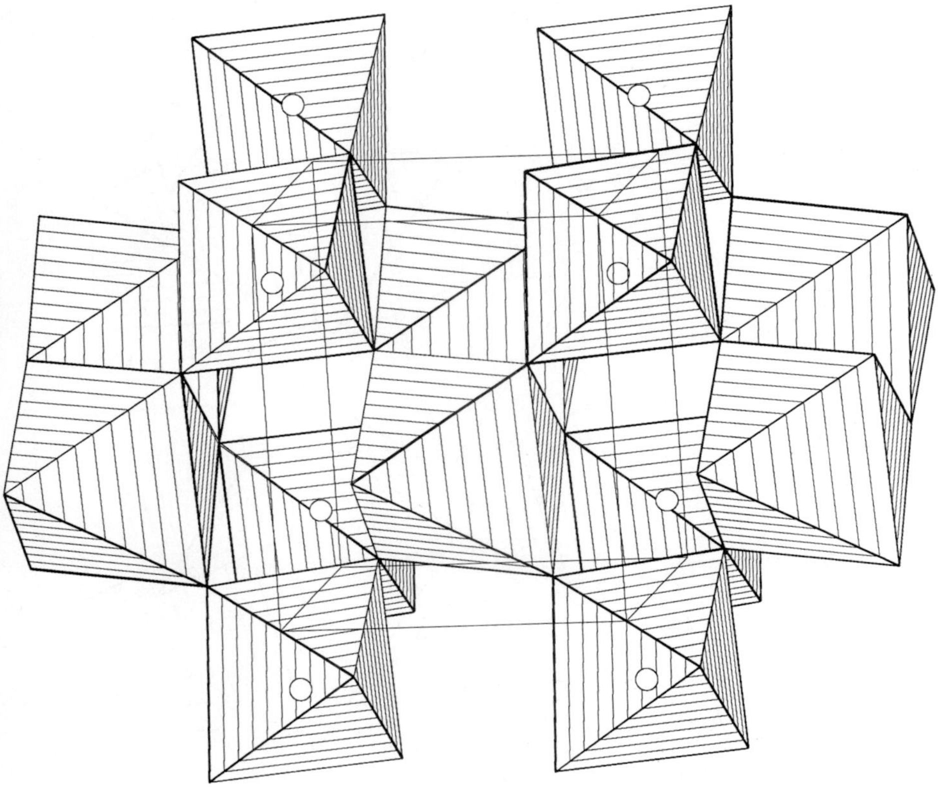

Figure 2.5a. Ferberite, $FeWO_4$, perspective c-axis projection, a horizontal. Both Fe and W (circle) are in octahedral coordination with point symmetry 2, although the W octahedron is more distorted. The manganese analogue huebnerite ($MnWO_4$) is isostructural.

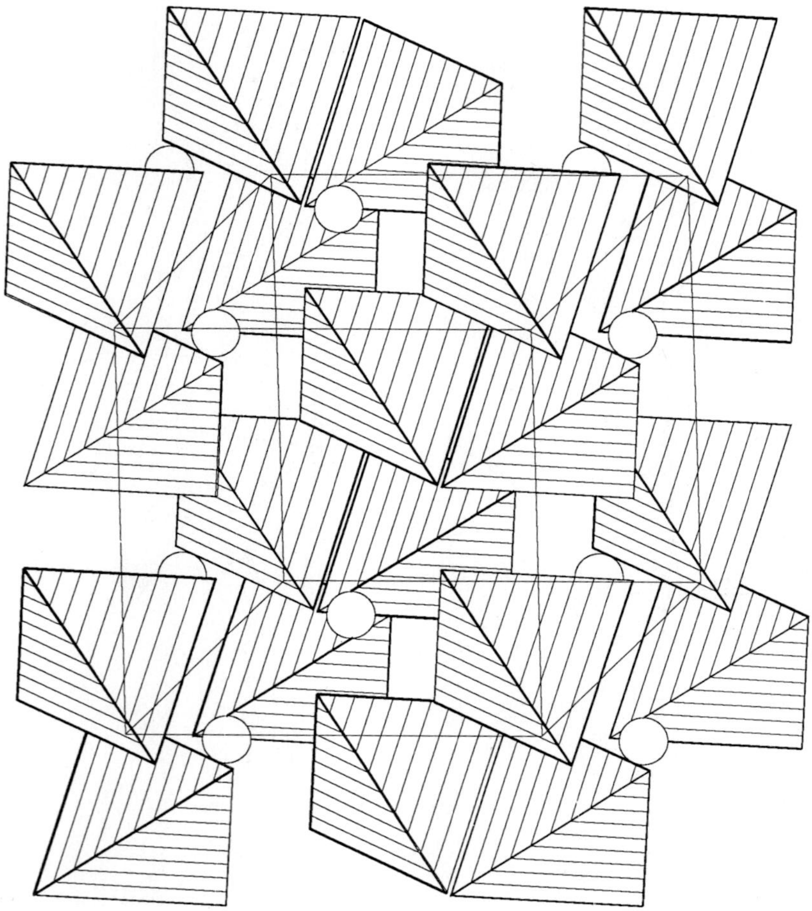

Figure 2.5b. Scheelite, $CaWO_4$, perspective c-axis projection, a horizontal. Ca (circle) is in eight-coordination with point symmetry $\bar{4}$ and, in contrast to ferberite (Fig 2.5a.), W is in tetrahedral coordination, also with point symmetry $\bar{4}$.

Table 2.5.2. Tungstate Divalent Cation Sites.

End-Member	Ferberite	Huebnerite	Scheelite
C.N.	6	6	8
Occupant	Fe^{+2}	Mn^{+2}	Ca
Point Sym.	2	2	$\bar{4}$
Wyckoff Not.	2f	2f	4b
Frac.Coord.			
x	1/2	1/2	0
y	.6744	.6866	1/4
z	1/4	1/4	5/8
Distances Cation-Ox			
1	2.057(2)	2.081(2)	2.479(4)
2	2.183(2)	2.294(2)	2.438(4)
3	2.146(2)	2.154(2)	
Mean	2.129	2.176	2.458
σ	.058	.097	.022
Poly.Vol.	12.559	13.286	26.376
Q.E.	1.0165	1.0246	-
Ang.Var.	56.06	80.63	-
Site Energy	-1297.	-1262.	-1144.

Table 2.5.3. Tungstate Tungsten Sites.

End-Member	Ferberite	Huebnerite	Scheelite
C.N.	6	6	4
Occupant	W^{+6}	W^{+6}	W^{+6}
Point Sym.	2	2	$\bar{4}$
Wyckoff Not.	2e	2e	4a
Frac.Coord.			
x	0	0	0
y	.1799	.1853	1/4
z	1/4	1/4	1/8
Distances Cation-Ox			
1	1.915(2)	1.936(2)	1.785(4)
2	2.122(2)	2.157(2)	
3	1.776(2)	1.756(2)	
Mean	1.938	1.950	1.785
σ	.156	.179	.000
Poly.Vol.	9.302	9.398	2.910
Q.E.	1.0339	1.0416	1.0024
Ang.Var.	95.94	115.87	9.65
Site Energy	-8265.	-8258.	-8225.

Chapter 3

HYDROXIDES

Table 3.1.1. Hydroxide Unit Cells

End-Member	Gibbsite	Diaspore	Boehmite	Brucite	Goethite
Formula	$Al(OH)_3$	$AlO(OH)$	$AlO(OH)$	$Mg(OH)_2$	$FeO(OH)$
Form.Wt.	78.004	59.988	59.988	58.327	88.854
Density	2.421	3.377	3.075	2.377	4.294
Mol Volume	32.222	17.862	19.507	24.524	20.693
Z	8	4	4	1	4
Cryst.Sys.	Monoclinic	Orthorhombic	Orthorhombic	Trigonal	Orthorhombic
Cryst.Class	$2/m$	*mmm*	*mmm*	3m1	mmm
Space Group	$P2_1/n$	*Pbnm*	*Amam*	$P\bar{3}m1$	*Pbnm*
Cell Parameters					
a	8.684	4.401	3.693	3.142	4.587
b	5.078	9.421	12.221	3.142	9.937
c	9.736	2.845	2.865	4.766	3.015
β	94.54				
Vol.	427.98	117.96	129.30	40.75	137.43
Ref.	Saalfeld & Wedde (1974)	Busing & Levy (1958)	Hill (1981)	Zigan & Rothbauer(1967)	Forsyth et al.(1968)

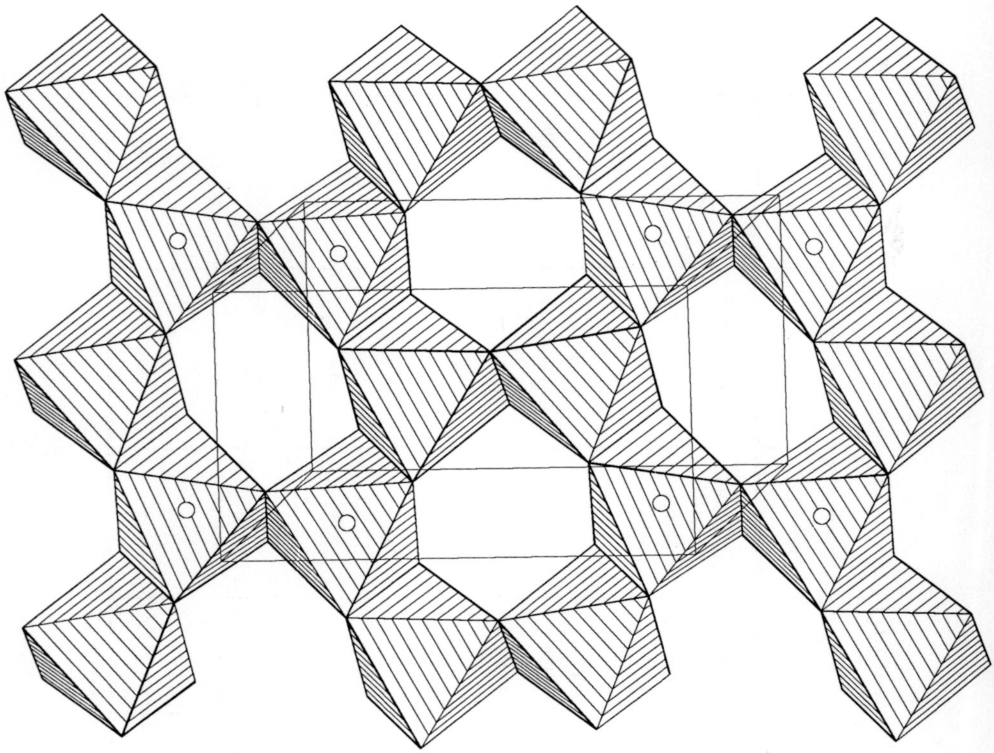

Figure 3.1. Gibbsite, $Al(OH)_3$, perspective c-axis projection, a horizontal. All anions are hydroxyls. There are two distinct Al sites, both with point symmetry 1. $Al(1)$ is indicated with small circles. This is the typical dioctahedral sheet.

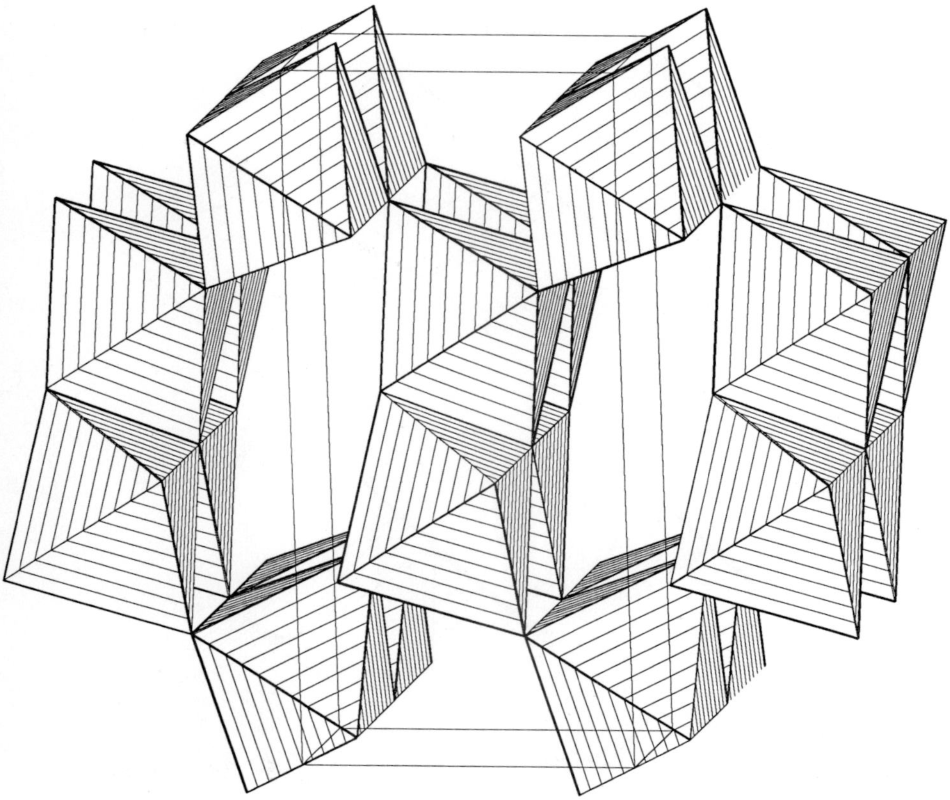

Figure 3.2. Diaspore, $AlO(OH)$, perspective c-axis projection, a horizontal. All Al sites are equivalent and are octahedrally coordinated with point symmetry m. Goethite ($FeO(OH)$) is isostructural.

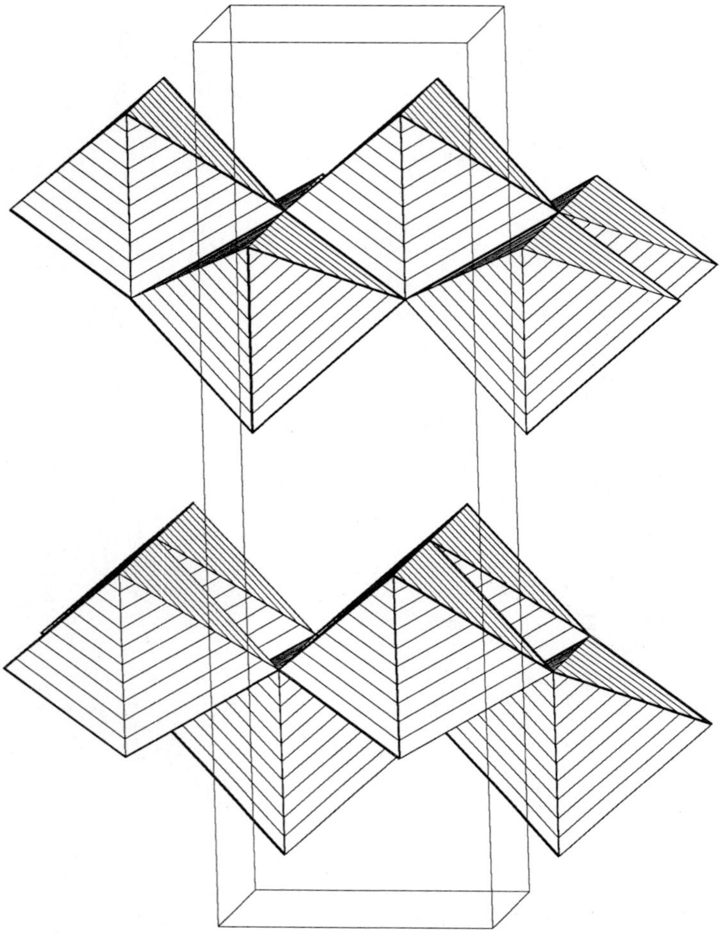

Figure 3.3. Boehmite, $AlO(OH)$, perspective c-axis projection, a horizontal. All Al sites are equivalent and octahedrally coordinated with point symmetry mm.

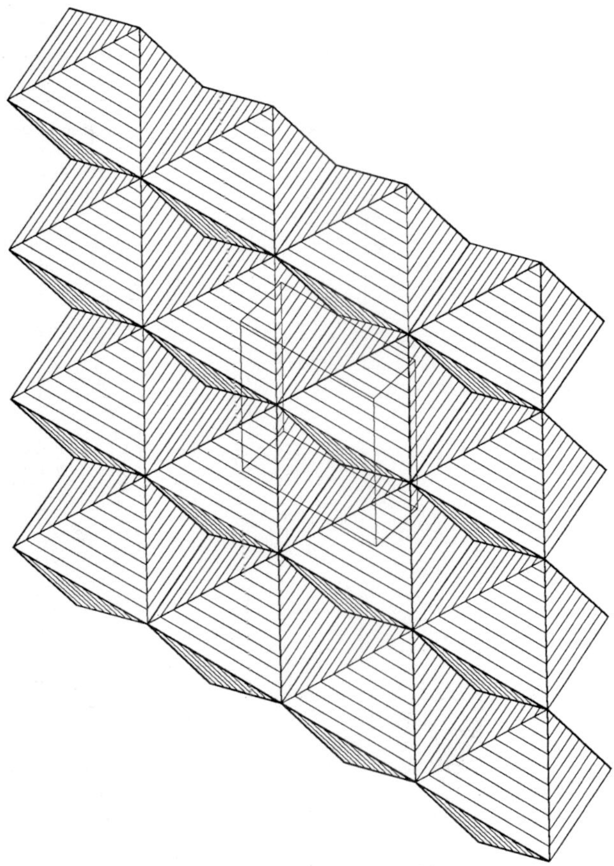

Figure 3.4. Brucite, $Mg(OH)_2$, perspective c-axis projection, [110] horizontal. All Mg sites are equivalent and octahedrally coordinated with point symmetry $\bar{3}m$. This is the typical trioctahedral sheet.

Table 3.1.2. Hydroxide Cation Sites

End-Member	Gibbsite	Gibbsite	Diaspore	Boehmite	Brucite	Goethite
C.N.	6	6	6	6	6	6
Occupant	Al(1)	Al(2)	Al	Al	Mg	Fe
Point Sym.	1	1	m	mm	$\bar{3}$m	m
Wyckoff Not.	4e	4e	4c	4c	1a	4c
Frac.Coord.						
x	0.1679	0.3344	0.0451	0.25	0	0.0477
y	0.5259	0.0236	-0.1446	-0.3178	0	0.8539
z	-0.0023	-0.0024	1/4	0	0	1/4
Distances						
O1	1.910	1.930	(2) 1.851 (1) 1.858	(2) 1.944 (2) 1.878 (2) 1.893	(6) 2.099	(2) 1.957 (1) 1.926
O2	1.922	1.922	(2) 1.975 (1) 1.980			(2) 2.095 (1) 2.095
O3	1.925	1.881				
O4	1.906	1.890				
O5	1.832	1.862				
O6	1.918	1.947				
Mean	1.902	1.905	1.915	1.905	2.099	2.021
σ	.035	.033	.067	.031	.000	.082
Poly.Vol.	9.019	9.061	9.100	9.000	12.039	10.667
Q.E.	1.0120	1.0120	1.0205	1.0164	1.0161	1.0226
Ang.Var.	42.5	42.5	64.0	52.0	52.0	69.2
Site Energy	-2408.	-2396.	-2484.	-2204.	-1139.	-2352.

Chapter 4

ORTHOSILICATES

Table 4.1.1. Garnet Group Unit Cells.

End-Member	Pyrope	Almandine	Spessartine	Grossular	Andradite	Uvarovite
Formula	$Mg_3Al_2Si_3O_{12}$	$Fe_3Al_2Si_3O_{12}$	$Mn_3Al_2Si_3O_{12}$	$Ca_3Al_2Si_3O_{12}$	$Ca_3Fe_2Si_3O_{12}$	$Ca_3Cr_2Si_3O_{12}$
Form.Wt.	403.150	497.755	495.028	450.454	508.185	500.483
Density	3.559	4.312	4.199	3.600	3.850	3.859
Mol Volume	113.28	115.43	117.88	125.12	131.99	129.71
Z	8	8	8	8	8	8
Cryst.Sys.	Isometric	Isometric	Isometric	Isometric	Isometric	Isometric
Cryst.Class	$m3m$	$m3m$	$m3m$	$m3m$	$m3m$	$m3m$
Space Group	$Ia3d$	$Ia3d$	$Ia3d$	$Ia3d$	$Ia3d$	$Ia3d$
Cell Parameters a	11.459	11.531	11.612	11.845	12.058	11.988
Vol.	1504.7	1533.2	1565.7	1661.9	1753.2	1722.8
Ref.	Novak & Gibbs (1971)	Novak & Gibbs (1971)	Novak & Gibbs (1971)	Novak & Gibbs (1971)	Novak & Gibbs (1971)	Novak & Gibbs (1971)

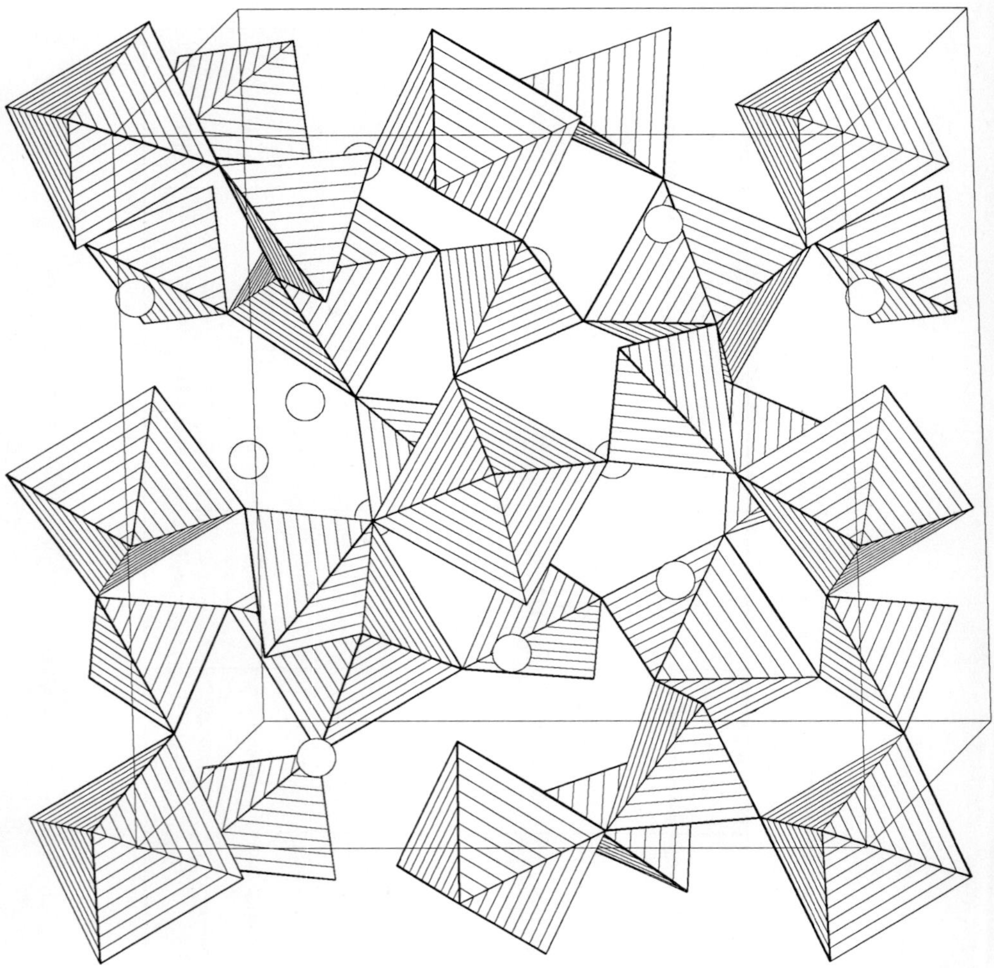

Figure 4.1. Pyrope, $Mg_3Al_2Si_3O_{12}$, perspective a-axis projection. Mg is in the X site (circle) in eight coordination with point symmetry 222. Al is in the Y site in octahedral coordination with point symmetry $\bar{3}$. Si is in the Z site in tetrahedral coordination with point symmetry $\bar{4}$. Although all $Si-O$ distances are identical, this Si site is one of the most distorted of all silicates.

Table 4.1.2. Garnet Group X-Sites.

End-Member	Pyrope	Almandine	Spessartine	Grossular	Andradite	Uvarovite
C.N.	8	8	8	8	8	8
Occupant	Mg	Fe	Mn	Ca	Ca	Ca
Point Sym.	222	222	222	222	222	222
Wyckoff Not.	24c	24c	24c	24c	24c	24c
Frac.Coord.						
x	1/8	1/8	1/8	1/8	1/8	1/8
y	0	0	0	0	0	0
z	1/4	1/4	1/4	1/4	1/4	1/4
Distances						
1(4)	2.197	2.220	2.245	2.319	2.365	2.360
2(4)	2.343	2.378	2.406	2.490	2.500	2.499
Mean	2.270	2.299	2.326	2.405	2.433	2.429
σ	.078	.084	.086	.091	.072	.075
Poly.Vol.	20.14	20.93	21.65	23.88	24.55	24.48
Site Energy	-1126.	-1101.	-1080.	-1022.	-1015.	-1010.

Table 4.1.3. Garnet Group Y-Sites.

End-Member	Pyrope	Almandine	Spessartine	Grossular	Andradite	Uvarovite
C.N.	6	6	6	6	6	6
Occupant	Al	Al	Al	Al	Fe^{+3}	Cr^{+3}
Point Sym.	$\bar{3}$	$\bar{3}$	$\bar{3}$	$\bar{3}$	$\bar{3}$	$\bar{3}$
Wyckoff Not.	16a	16a	16a	16a	16a	16a
Frac.Coord.						
x	0	0	0	0	0	0
y	0	0	0	0	0	0
z	0	0	0	0	0	0
Distances						
O(6)	1.887	1.896	1.901	1.924	2.024	1.985
Poly.Vol.	8.937	9.086	9.155	9.491	11.046	10.413
Q.E.	1.0014	1.0004	1.0001	1.0007	1.0004	1.0007
Ang.Var.	4.93	1.38	0.30	2.33	1.36	2.64
Site Energy	-2666.	-2655.	-2658.	-2640.	-2455.	-2527.

Table 4.1.4. Garnet Group Z-Sites.

End-Member	Pyrope	Almandine	Spessartine	Grossular	Andradite	Uvarovite
C.N.	4	4	4	4	4	4
Occupant	Si	Si	Si	Si	Si	Si
Point Sym.	$\bar{4}$	$\bar{4}$	$\bar{4}$	$\bar{4}$	$\bar{4}$	$\bar{4}$
Wyckoff Not.	24d	24d	24d	24d	24d	24d
Frac.Coord.						
x	3/8	3/8	3/8	3/8	3/8	3/8
y	0	0	0	0	0	0
z	1/4	1/4	1/4	1/4	1/4	1/4
Distances						
O(4)	1.635	1.628	1.636	1.645	1.643	1.643
Poly.Vol.	2.192	2.172	2.206	2.261	2.250	2.257
Q.E.	1.0150	1.0134	1.0117	1.0073	1.0071	1.0058
Ang.Var.	61.6	55.2	48.0	29.8	28.9	23.9
Site Energy	-4380.	-4413.	-4400.	-4402.	-4432.	-4426.

Table 4.2.1. Olivine Group Unit Cells.

End-Member	Forsterite	Fayalite	Monticellite	Kirschsteinite	Ca-Olivine	Tephroite	Co-Olivine	Liebenbergite
Formula	Mg_2SiO_4	Fe_2SiO_4	$MgCaSiO_4$	$FeCaSiO_4$	Ca_2SiO_4	Mn_2SiO_4	Co_2SiO_4	Ni_2SiO_4
Form.Wt.	140.708	203.778	156.476	188.011	172.244	201.960	209.950	209.503
Density	3.227	4.402	3.040	3.965	2.969	4.127	4.719	4.921
Mol Volume	43.603	46.290	51.472	47.415	58.020	48.939	44.493	42.574
Z	4	4	4	4	4	4	4	4
Cryst.Sys.	Orthorhombic	Orthorhombic	Orthorhombic	Orthorhombic	Orthorhombic	Orthorhombic	Orthorhombic	Orthorhombic
Cryst.Class	*mmm*	*mmm*	*mmm*	*mmm*	*mmm*	*mmm*	*mmm*	*mmm*
Space Group	*Pbnm*	*Pbnm*	*Pbnm*	*Pbnm*	*Pbnm*	*Pbnm*	*Pbnm*	*Pbnm*
Cell Parameters								
a	4.7534	4.8195	4.822	4.844	5.078	4.9023	4.7811	4.726
b	10.1902	10.4788	11.108	10.577	11.225	10.5964	10.2998	10.118
c	5.9783	6.0873	6.382	6.146	6.760	6.2567	6.0004	5.913
Vol.	289.58	307.42	341.84	314.89	385.32	325.02	295.49	282.75
Ref.	Fujino et al.(1981)	Fujino et al.(1981)	Onken (1965)	Brown (1970)	Czaya (1971)	Fujino et al.(1981)	Brown (1970)	Lager & Meagher(1978)

Figure 4.2. Forsterite, Mg_2SiO_4, perspective c-axis projection, a horizontal. There are two distinct Mg octahedra, M1 and M2 (circle), and a single Si tetrahedron. M1 is smaller and slightly more distorted than M2. M1 has point symmetry $\bar{1}$, and M2 and Si lie on the mirror plane with point symmetry m.

Table 4.2.2. Olivine M1 Site.

End-Member	Forsterite	Fayalite	Monticellite	Kirschsteinite	Ca-Olivine	Tephroite	Co-Olivine	Liebenbergite
C.N.	6	6	6	6	6	6	6	6
Occupant	Mg	Fe^{+2}	Mg	Fe^{+2}	Ca	Mn	Co	Ni
Pt.Sym.	$\bar{1}$	$\bar{1}$	$\bar{1}$	$\bar{1}$	$\bar{1}$	$\bar{1}$	$\bar{1}$	$\bar{1}$
Wyckoff Not.	4a	4a	4a	4a	4a	4a	4a	4a
Frac. Coords.								
x	0	0	0	0	0	0	0	0
y	0	0	0	0	0	0	0	0
z	0	0	0	0	0	0	0	0
Distances								
O1(2)	2.0838	2.1207	2.193	2.139	2.356	2.2003	2.098	2.064
O2(2)	2.0678	2.1259	2.090	2.098	2.311	2.1671	2.091	2.060
O3(2)	2.1311	2.2363	2.119	2.154	2.388	2.2498	2.167	2.111
Mean	2.094	2.161	2.134	2.130	2.352	2.206	2.119	2.078
σ	.029	.058	.047	.026	.035	.037	.037	.026
Poly. Vol.	11.771	12.737	12.420	12.105	15.896	13.499	12.144	11.531
Q.E.	1.0269	1.0379	1.0287	1.0427	1.0601	1.0398	1.0294	1.0254
Ang.Var.	95.3	130.1	100.3	147.9	209.0	138.8	102.9	90.3
Site Energy	-1082.	-1041.	-1078.	-1063.	-945.	-1015.	-1064.	-1090.

Table 4.2.3. Olivine M2 Sites.

End-Member	Forsterite	Fayalite	Monticellite	Kirschsteinite	Ca-Olivine	Tephroite	Co-Olivine	Liebenbergite
C.N.	6	6	6	6	6	6	6	6
Occupant	Mg	Fe^{+2}	Ca	Ca	Ca	Mn	Co	Ni
Pt.Sym.	*m*	*m*	*m*	*m*	*m*	*m*	*m*	*m*
Wyckoff Not.	4c	4c	4c	4c	4c	4c	4c	4c
Frac. Coords.								
x	.99169	.98598	.9770	.9888	.9904	.98792	.9915	.9924
y	.27739	.28026	.2767	.2799	.2809	.28041	.2764	.2738
z	1/4	1/4	1/4	1/4	1/4	1/4	1/4	1/4
Distances								
O1(1)	2.1766	2.2331	2.478	2.388	2.441	2.2782	2.187	2.105
O2(1)	2.0454	2.1109	2.308	2.146	2.286	2.1369	2.072	2.043
O3(2)	2.0658	2.0647	2.287	2.212	2.385	2.1547	2.073	2.053
O3(2)	2.2101	2.2946	2.411	2.325	2.426	2.3194	2.223	2.171
Mean	2.129	2.177	2.364	2.268	2.392	2.227	2.142	2.100
σ	.078	.110	.080	.092	.057	.088	.077	.066
Poly. Vol.	12.401	13.072	16.438	14.549	16.930	13.982	12.606	11.966
Q.E.	1.0260	1.0370	1.0481	1.0468	1.0516	1.0367	1.0269	1.0215
Ang.Var.	89.5	124.9	165.6	161.1	180.8	127.0	92.7	74.9
Site Energy	-1160.	-1136.	-1010.	-1061.	-1008.	-1102.	-1153.	-1179.

Table 4.2.4. Olivine Tetrahedral Sites.

End-Member	Forsterite	Fayalite	Monticellite	Kirschsteinite	Ca-Olivine	Tephroite	Co-Olivine	Liebenbergite
C.N.	4	4	4	4	4	4	4	4
Occupant	Si	Si	Si	Si	Si	Si	Si	Si
Pt.Sym.	*m*	*m*	*m*	*m*	*m*	*m*	*m*	*m*
Wyckoff Not.	4c	4c	4c	4c	4c	4c	4c	4c
Frac. Coords.								
x	.42645	.43122	.4101	.4181	.4283	.42755	.4282	.4276
y	.09403	.09765	.0811	.0846	.0959	.09643	.0949	.0944
z	1/4	1/4	1/4	1/4	1/4	1/4	1/4	1/4
Distances								
O1(1)	1.6139	1.6248	1.614	1.612	1.633	1.6191	1.613	1.620
O2(1)	1.6549	1.6533	1.656	1.568	1.655	1.6578	1.659	1.660
O3(2)	1.6368	1.6333	1.639	1.551	1.647	1.6395	1.636	1.637
Mean	1.636	1.636	1.637	1.570	1.646	1.639	1.636	1.638
σ	.017	.012	.017	.029	.009	.016	.019	.016
Poly. Vol.	2.209	2.220	2.220	1.958	2.266	2.232	2.216	2.218
Q.E.	1.0112	1.0085	1.0092	1.0102	1.0062	1.0082	1.0100	1.0118
Ang.Var.	49.4	36.7	40.6	36.8	27.2	36.1	44.3	52.3
Site Energy	-4319.	-4348.	-4333.	-4551.	-4377.	-4349.	-4326.	-4305.

Table 4.3.1. Silicate Spinel Unit Cells.

End-Member	Mg_2SiO_4	Fe_2SiO_4	Co_2SiO_4	Ni_2SiO_4
Form.Wt.	140.708	203.778	209.950	209.504
Density	3.563	4.848	5.174	5.346
Mol Volume	39.493	42.030	40.577	39.187
Z	8	8	8	8
Cryst.Sys.	Isometric	Isometric	Isometric	Isometric
Cryst.Class	$m3m$	$m3m$	$m3m$	$m3m$
Space Group	$Fd3m$	$Fd3m$	$Fd3m$	$Fd3m$
Cell Parameters				
a	8.0649	8.234	8.138	8.044
Vol.	524.56	558.26	538.96	520.49
Ref.	Sasaki et al.(1982a)	Yagi et al.(1974)	Morimoto et al.(1974)	Yagi et al.(1974)

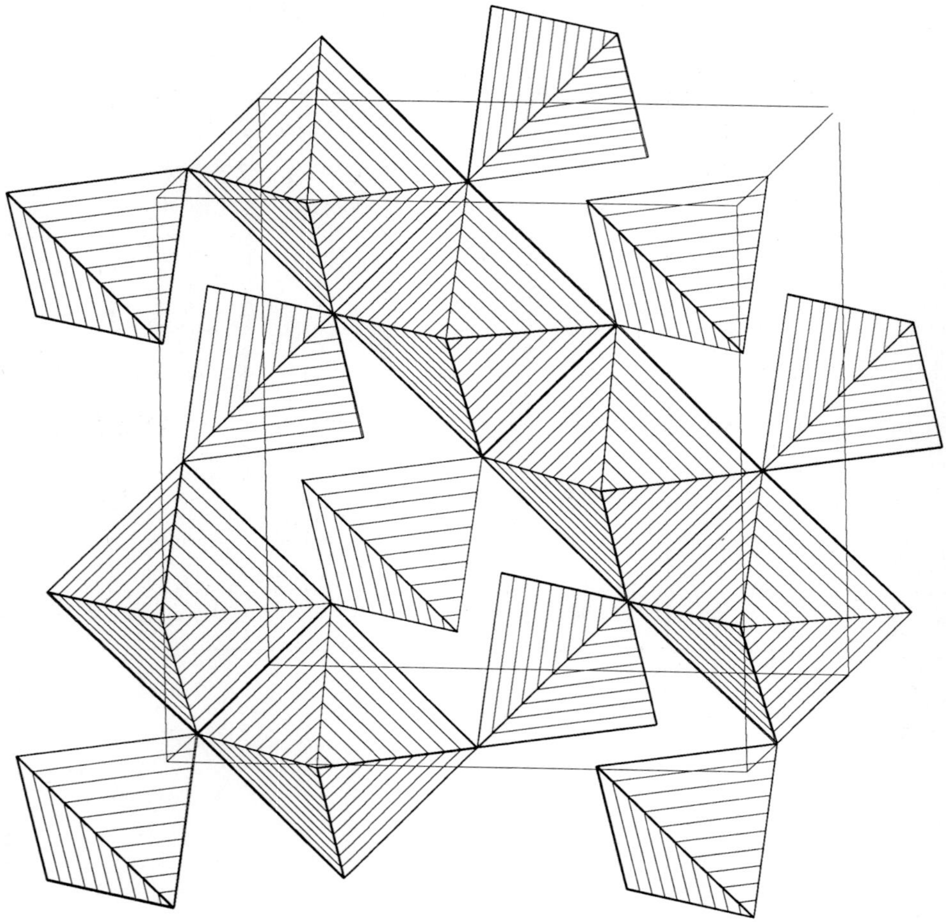

Figure 4.3. Silicate spinel, Mg_2SiO_4, perspective a-axis projection. Mg is in octahedral coordination with point symmetry $\bar{3}m$, and Si is in tetrahedral coordination with point symmetry $\bar{4}3m$. In contrast to olivine (Figure 4.2) and the beta orthosilicate structure (Figure 5.3) with which this compound is polymorphous, there is a single Mg site in this structure.

Table 4.3.2. Silicate Spinel Octahedral Sites.

End-Member	Mg_2SiO_4	Fe_2SiO_4	Co_2SiO_4	Ni_2SiO_4
C.N.	6	6	6	6
Occupant	Mg	Fe	Co	Ni
Point Sym.	$\bar{3}$m	$\bar{3}$m	$\bar{3}$m	$\bar{3}$m
Wyckoff Not.	16d	16d	16d	16d
Frac.Coord.				
x	1/2	1/2	1/2	1/2
y	1/2	1/2	1/2	1/2
z	1/2	1/2	1/2	1/2
Distances				
O (6)	2.070	2.137	2.103	2.063
Poly.Vol.	11.780	12.912	12.332	11.663
Q.E.	1.0026	1.0051	1.0041	1.0024
Ang.Var.	8.95	17.35	13.95	8.43
Site Energy	-1155.	-1101.	-1125.	-1160.

Table 4.3.3. Silicate Spinel Tetrahedral Sites.

End-Member	Mg_2SiO_4	Fe_2SiO_4	Co_2SiO_4	Ni_2SiO_4
C.N.	4	4	4	4
Occupant	Si	Si	Si	Si
Point Sym.	$\bar{4}3m$	$\bar{4}3m$	$\bar{4}3m$	$\bar{4}3m$
Wyckoff Not.	8a	8a	8a	8a
Frac.Coord.				
x	1/8	1/8	1/8	1/8
y	1/8	1/8	1/8	1/8
z	1/8	1/8	1/8	1/8
Distances				
O (4)	1.655	1.652	1.646	1.654
Poly.Vol.	2.328	2.312	2.290	2.321
Q.E.	1.0000	1.0000	1.0000	1.0000
Ang.Var.	0.00	0.00	0.00	0.00
Site Energy	-4417.	-4459.	-4461.	-4419.

Table 4.4.1. Silicate Zircon Unit Cells.

End-member	Zircon	Hafnon	Thorite	Coffinite
Formula	$ZrSiO_4$	$HfSiO_4$	$ThSiO_4$	$USiO_4$
Form.Wt.	183.304	270.574	324.122	330.114
Density	4.668	6.976	6.696	7.185
Mol Volume	39.270	38.787	48.407	45.945
Z	4	4	4	4
Cryst.Sys.	Tetragonal	Tetragonal	Tetragonal	Tetragonal
Cryst.Class	$4mm$	$4mm$	$4mm$	$4mm$
Space Group	$I4_1/amd$	$I4_1/amd$	$I4_1/amd$	$I4_1/amd$
Cell Parameters				
a	6.6042	6.5725	7.1328	6.995
c	5.9796	5.9632	6.3188	6.236
Vol.	260.80	257.60	321.48	305.13
Ref.	Hazen & Finger(1979)	Speer & Cooper(1982)	Taylor & Ewing(1978)	Keller (1963)

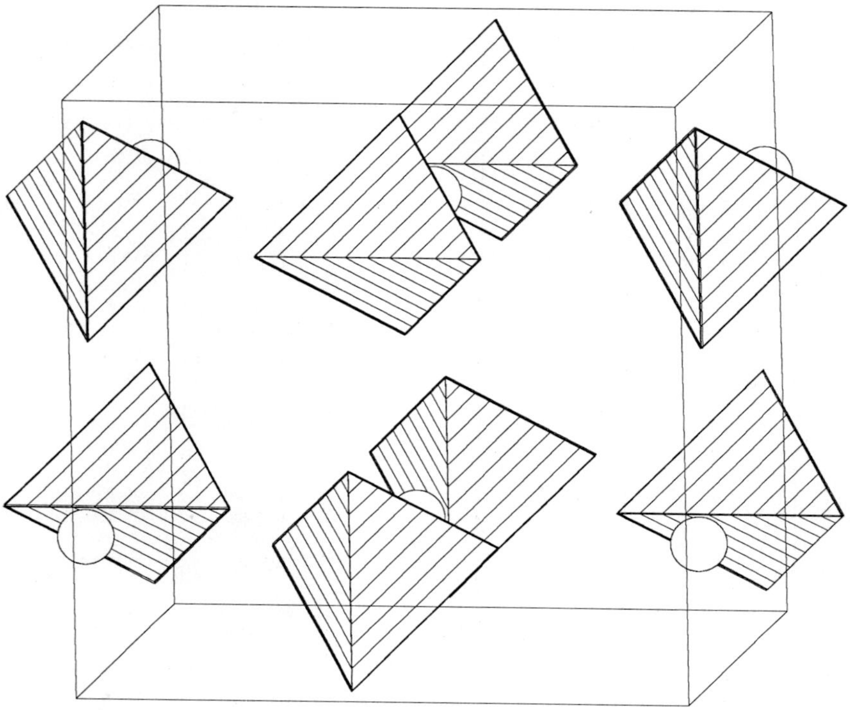

Figure 4.4. Zircon, $ZrSiO_4$, perspective c-axis projection. Zr (circle) is in eight-coordination, and Si is tetrahedral. Both sites have point symmetry $\bar{4}2m$.

Table 4.4.2. Silicate Zircon M Sites.

End-Member	Zircon	Hafnon	Thorite	Coffinite
C.N.	8	8	8	8
Occupant	Zr	Hf	Th	U
Point Sym.	$\bar{4}2m$	$\bar{4}2m$	$\bar{4}2m$	$\bar{4}2m$
Wyckoff Not.	4a	4a	4a	4a
Frac.Coord.				
x	0	0	0	0
y	3/4	3/4	3/4	3/4
z	1/8	1/8	1/8	1/8
Distances				
1(4)	2.129	2.115	2.368	2.323
2(4)	2.267	2.260	2.466	2.430
Mean	2.198	2.187	2.417	2.376
σ	0.074	0.077	0.053	0.057
Poly.Vol.	19.00	18.72	25.32	24.04
Site Energy	-3884.	-3906.	-3455.	-3518.

Table 4.4.3. Silicate Zircon Tetrahedral Sites.

End-Member	Zircon	Hafnon	Thorite	Coffinite
C.N.	4	4	4	4
Occupant	Si	Si	Si	Si
Point Sym.	$\bar{4}2m$	$\bar{4}2m$	$\bar{4}2m$	$\bar{4}2m$
Wyckoff Not.	4b	4b	4b	4b
Frac.Coord.				
x	0	0	0	0
y	1/4	1/4	1/4	1/4
z	3/8	3/8	3/8	3/8
Distances				
1(4)	1.623	1.620	1.635	1.607
Poly.Vol.	2.118	2.107	2.205	2.093
Q.E.	1.0237	1.0239	1.0109	1.0118
Ang. Var.	97.8	98.7	44.8	48.6
Site Energy	-4513.	-4519.	-4545.	-4626.

Table 4.5.1. Willemite Group Unit Cells.

End-member	Willemite	Phenacite
Formula	Zn_2SiO_4	Be_2SiO_4
Form.Wt.	222.824	110.108
Density	4.221	2.960
Mol Volume	52.795	37.197
Z	18	18
Cryst.Sys.	Trigonal	Trigonal
Cryst.Class	$\bar{3}$	$\bar{3}$
Space Group	$R\bar{3}$	$R\bar{3}$
Cell Parameters		
a	13.971	12.472
c	9.334	8.252
Vol.	1577.8	1111.6
Ref.	Simonov (1977)	Zachariasen (1971)

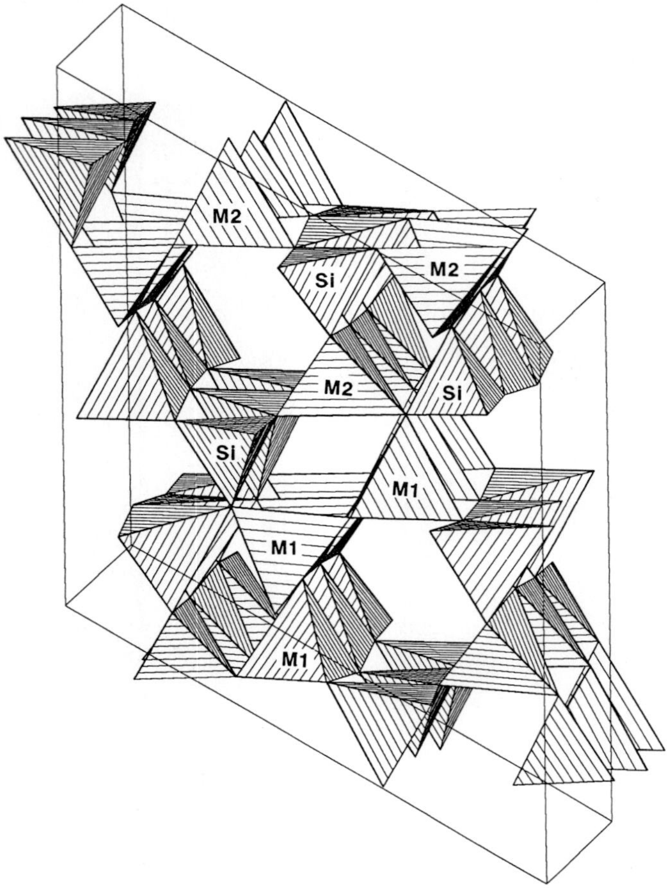

Figure 4.5. Willemite, Zn_2SiO_4, perspective c-axis projection. There are two Zn sites, M1 and M2, and a single Si site, all with point symmetry 1. Phenacite (Be_2SiO_4) is isostructural. Although all cations are tetrahedral, this is not a framework silicate because the coordination number of oxygen is three (one Si and two Zn) rather than two as in the framework silicates.

Table 4.5.2. Willemite Group M-sites.

End-Member	Willemite	Phenacite		Willemite	Phenacite
Site	M1	M1		M2	M2
C.N.	4	4		4	4
Occupant	Zn	Be		Zn	Be
Point Sym.	1	1		1	1
Wyckoff Not.	18f	18f		18f	18f
Frac.Coord.					
x	.2087	.19397		.2155	.19386
y	.0171	.98412		.0234	.98234
z	.4156	.41547		.0815	.08454
Distances					
O1	1.958	1.640	O1	1.958	1.631
O2	1.952	1.645	O2	1.967	1.643
O4	1.965	1.658	O3	1.972	1.655
O4	1.957	1.637	O3	2.008	1.655
Mean	1.958	1.645		1.976	1.646
σ	.005	.009		.022	.011
Poly.Vol.	3.821	2.280		3.934	2.283
Q.E.	1.0054	1.0014		1.0048	1.0017
Ang.Var.	21.43	5.38		19.63	7.19
Site Energy	-1123.	-1379.		-1109.	-1397.

Table 4.5.3. Willemite Group Si-sites.

End-Member	Willemite	Phenacite
C.N.	4	4
Occupant	Si	Si
Point Sym.	1	1
Wyckoff Not.	18f	18f
Frac.Coord.		
x	.2118	.19559
y	.0155	.98402
z	.7490	.74993
Distances		
O1	1.626	1.630
O2	1.611	1.628
O3	1.637	1.634
O4	1.619	1.631
Mean	1.623	1.631
σ	.011	.002
Poly.Vol.	2.191	2.222
Q.E.	1.0009	1.0009
Ang.Var.	3.29	3.81
Site Energy	-4445.	-4338.

Table 4.6.1. Aluminosilicate Unit Cells.

Polymorph	Andalusite	Sillimanite	Kyanite	Topaz
Formula	Al_2SiO_5	Al_2SiO_5	Al_2SiO_5	$Al_2SiO_4(OH,F)_2$
Form.Wt.	162.046	162.046	162.046	182.052
Density	3.1426	3.2386	3.6640	3.492
Mol Volume	51.564	50.035	44.227	52.140
Z	4	4	4	4
Cryst.Sys.	Orthorhombic	Orthorhombic	Triclinic	Orthorhombic
Cryst.Class	*mmm*	*mmm*	$\bar{1}$	*mmm*
Space Group	*Pnnm*	*Pbnm*	$P\bar{1}$	*Pbnm*
Cell Parameters				
a	7.7980	7.4883	7.1262	4.6651
b	7.9031	7.6808	7.8520	8.8381
c	5.5566	5.7774	5.5747	8.3984
α			89.99	
β			101.11	
γ			106.03	
Vol.	342.44	332.29	293.72	346.27
Ref.	Winter & Ghose(1979)	Winter & Ghose(1979)	Winter & Ghose(1979)	Zemann et al.(1979)

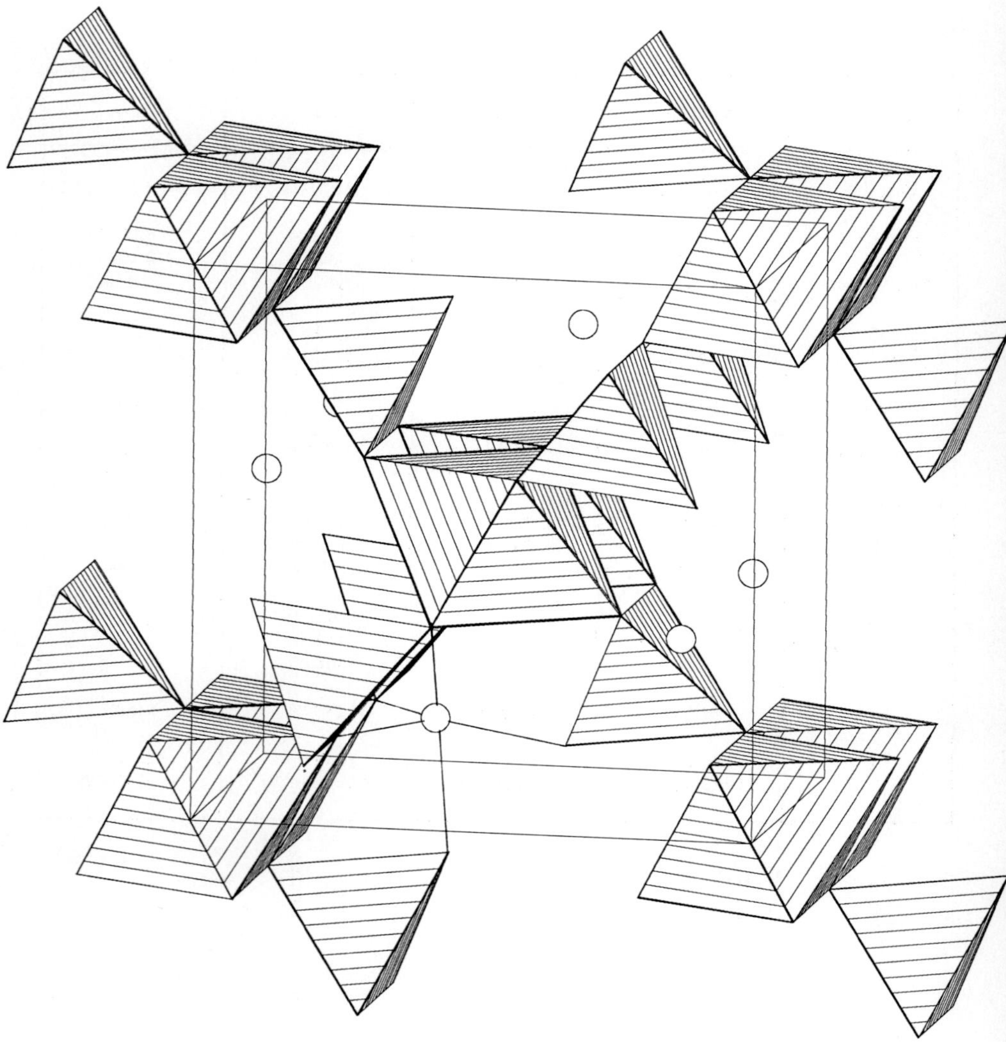

Figure 4.6a. Andalusite, Al_2SiO_5, perspective c-axis projection. Al1 is in octahedral coordination with point symmetry 2, whereas Al2 (circle) is in five-coordination with point symmetry m. There is a single Si site in tetrahedral coordination with point symmetry m.

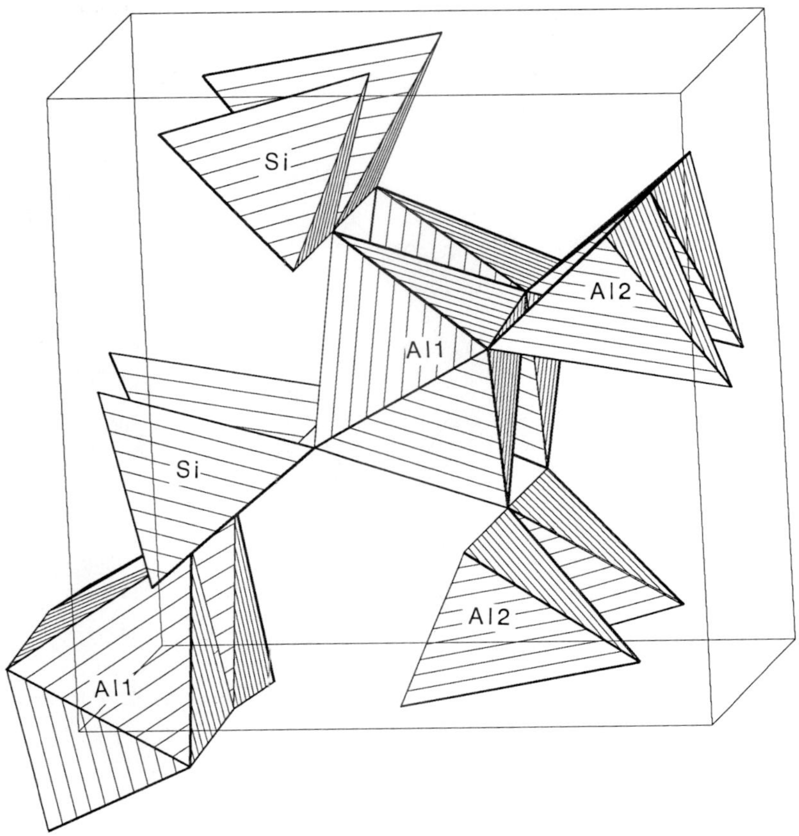

Figure 4.6b. Sillimanite, Al_2SiO_5, perspective c-axis projection, a horizontal. There are two distinct Al sites, both in octahedral coordination. Al1 has point symmetry $\bar{1}$, and Al2 has point symmetry m. There is a single Si site in tetrahedral coordination and point symmetry m.

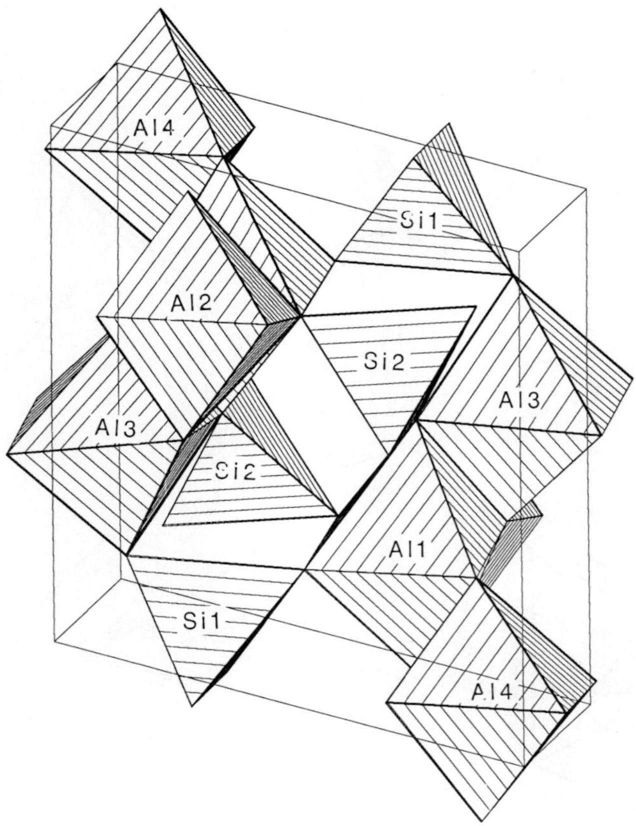

Figure 4.6c. Kyanite, Al_2SiO_5, perspective c-axis projection, b vertical. There are four distinct Al octahedra and two distinct Si tetrahedra, all in the general position with point symmetry 1.

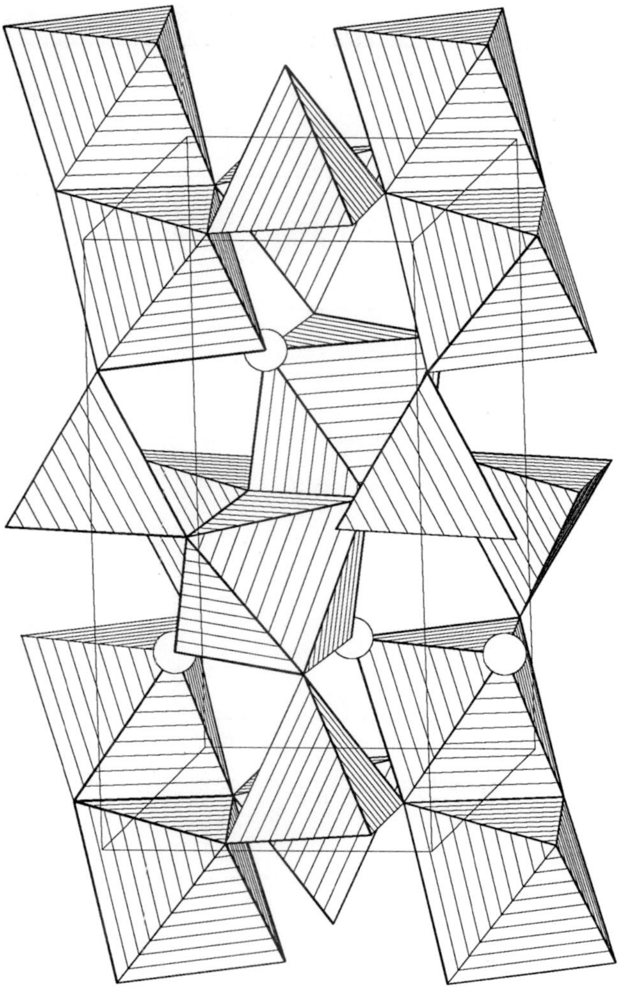

Figure 4.6d. Topaz, $Al_2SiO_4(OH,F)_2$, perspective c-axis projection, b vertical. There is a single Si site in tetrahedral coordination and a single Al in octahedral coordination. The F (circle) bonds only to Al.

Table 4.6.2. Aluminosilicate Aluminum Sites.

End-Member	Andalusite		Sillimanite		Kyanite				Topaz
Site	Al1	Al2	Al1	Al2	Al1	Al2	Al3	Al4	Al
C.N.	6	5	6	4	6	6	6	6	6
Occupant	Al	Al	Al	Al	Al	Al	Al	Al	Al
Point Sym.	2	m	$\bar{1}$	m	1	1	1	1	1
Wyckoff Not.	4e	4g	4a	4c	2i	2i	2i	2i	8d
Frac.Coord.									
x	0	.3705	0	.1417	.3254	.2974	.0998	.1120	.90516
y	0	.1391	0	.3449	.7040	.6989	.3862	.9175	.13123
z	.2419	1/2	0	1/4	.4582	.9505	.6403	.1649	.08180
Distances	(OA) 1.827 (2) (OB) 1.891 (2) (OD) 2.086 (2)	(OA) 1.816 (OC) 1.840 (OC) 1.899 (OD) 1.814 (2)	(OA) 1.914 (2) (OB) 1.868 (2) (OD) 1.954 (2)	(OB) 1.751 (OC) 1.711 (OD) 1.796 (2)	(OB) 1.874 (OF) 1.884 (OG) 1.971 (OH) 1.987 (OK) 1.847 (OM) 1.848	(OB) 1.934 (OC) 1.881 (OD) 1.889 (OF) 1.914 (OK) 1.930 (OM) 1.925	(OB) 1.986 (OC) 1.924 (OE) 1.862 (OF) 1.968 (OF) 1.883 (OG) 1.885	(OA) 1.816 (OA) 1.998 (OB) 1.846 (OD) 1.911 (OE) 1.933 (OH) 1.875	(O1)1.908 (O2)1.911 (O3)1.894 (O3')1.902 (F) 1.808 (F')1.802
Mean	1.935	1.836	1.912	1.764	1.902	1.913	1.918	1.896	1.871
σ	.121	.036	.039	.041	.062	.023	.050	.065	.051
Poly.Vol.	9.539	5.153	9.175	2.791	8.977	9.136	9.164	8.921	8.654
Q.E.	1.0114	--	1.0109	1.0062	1.0155	1.0141	1.0180	1.0139	1.0066
Ang. Var.	18.0	--	36.4	20.5	47.7	50.2	57.0	42.5	20.6
Site Energy	-2490.	-2569.	-2573.	-2526.	-2532.	-2563.	-2543.	-2531.	-2506.

Table 4.6.3. Aluminosilicate Group Si Sites.

End-Member	Anadalusite	Sillimanite	Kyanite		Topaz
Site	Si	Si	Si1	Si2	Si
C.N.	4	4	4	4	4
Occupant	Si	Si	Si	Si	Si
Point Sym.	m	m	1	1	m
Wyckoff Not.	4g	4c	2i	2i	4c
Frac.Coord.					
x	.2460	.1533	.2692	.2910	.39955
y	.2520	.3402	.0649	.3317	.94084
z	0	3/4	.7066	.1892	1/4
Distances					
1	(OB) 1.646	(OA) 1.640	(OD) 1.631	(OA) 1.640	(O1)1.637
2	(OC) 1.618	(OC) 1.573	(OE) 1.643	(OC) 1.629	(O2)1.651
	(OD) 1.630 (2)	(OD) 1.645 (2)	(OH) 1.621	(OG) 1.627	(O3)1.643(2)
			(OM) 1.647	(OK) 1.649	
Mean	1.631	1.626	1.636	1.636	1.643
σ	.011	.035	.011	.010	.006
Poly.Vol.	2.211	2.203	2.241	2.243	2.277
Q.E.	1.0043	1.0013	1.0012	1.0018	1.0004
Ang. Var.	16.4	3.4	4.8	7.1	1.7
Site Energy	-4404.	-4426.	-4443.	-4458.	-4402.

Table 4.7.1. Humite Group Unit Cells.

End-Member	Norbergite	Chondrodite	Humite	Clinohumite
Formula	Mg_3 (SiO_4) $F_{1.8}(OH)_{.2}$	$(Mg_{4.95}Fe_{.05})$ $(SiO_4)_2$ $F_{1.3}(OH)_{.7}$	$(Mg_{6.6}Fe_{.4})$ $(SiO_4)_3$ $F(OH)$	$(Mg_{8.42}Fe_{.56})$ $(SiO_4)_4$ $F_{1.04}(OH)_{.93}$
Form.Wt.	203.016	341.73	482.44	640.49
Density	3.186	3.158	3.159	3.259
Mol Volume	63.73	108.20	152.70	196.55
Z	4	2	4	2
Cryst.Sys.	Orthorhombic	Monoclinic	Orthorhombic	Monoclinic
Cryst.Class	*mmm*	$2/m$	*mmm*	$2/m$
Space Group	*Pbnm*	$P2_1/b$	*Pbnm*	$P2_1/b$
Cell Parameters				
a	4.7104	4.7284	4.7408	4.7441
b	10.2718	10.2539	10.2580	10.2501
c	8.7476	7.8404	20.8526	13.6635
α		109.059		100.786
Vol.	423.25	359.30	1014.09	652.68
Ref.	Gibbs & Ribbe (1969)	Gibbs et al. (1970)	Ribbe & Gibbs (1971)	Robinson et al. (1973a)

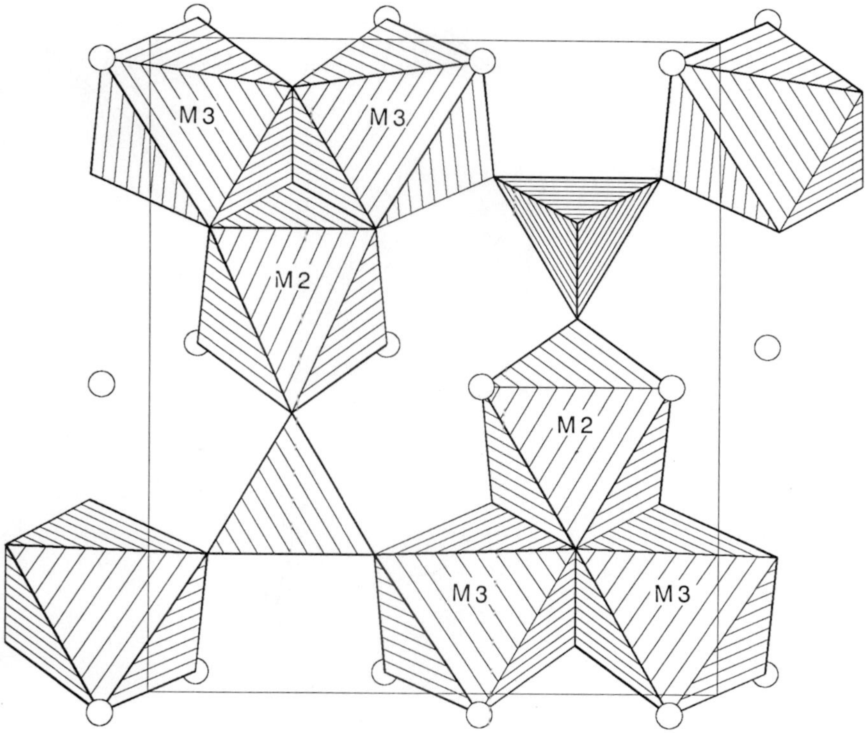

Figure 4.7a. Norbergite, $Mg_3SiO_4(F,OH)_2$, a-axis projection, b vertical. There are two distinct Mg octahedra, M2 (point symmetry m) and M3 (point symmetry 1) and a single Si tetrahedron. The F (small circle) bonds only to Mg.

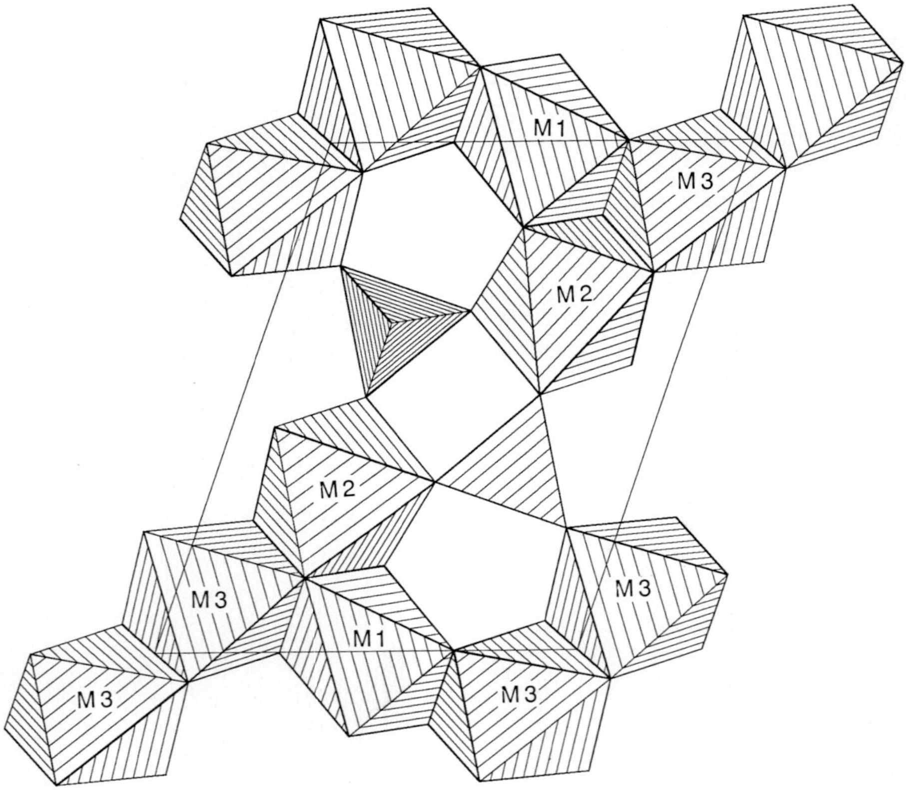

Figure 4.7b. Chondrodite, $Mg_5(SiO_4)_2(F,OH)_2$, a-axis projection, c horizontal. There are three distinct Mg octahedra and a single Si tetrahedron.

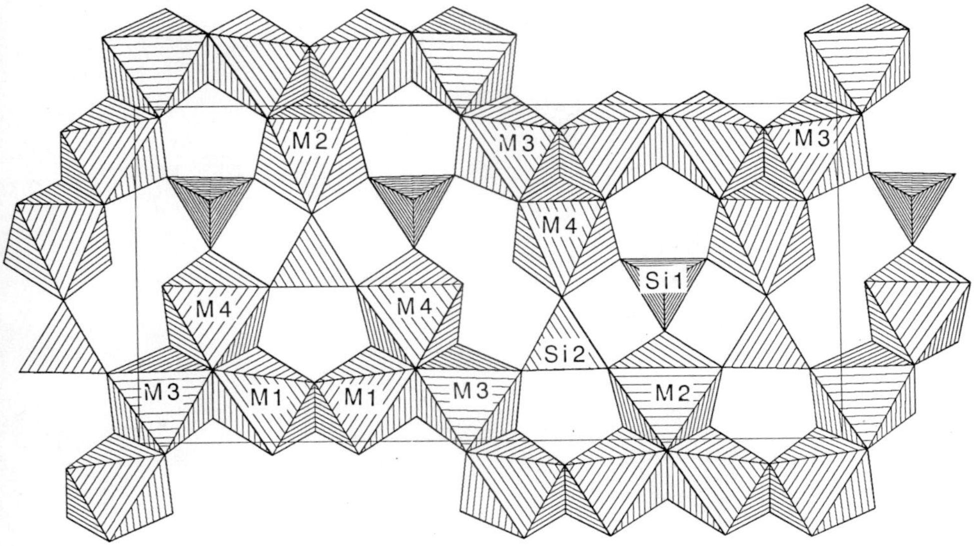

Figure 4.7c. Humite, $(Mg,Fe)_7(SiO_4)_3(F,OH)_2$, a-axis projection, b vertical. There are four distinct octahedral sites occupied by Mg and Fe; M1, M3, and M4 have point symmetry 1, and M2 has point symmetry m. There are two distinct tetrahedral sites; Si1 with point symmetry m, and Si2 with point symmetry 1.

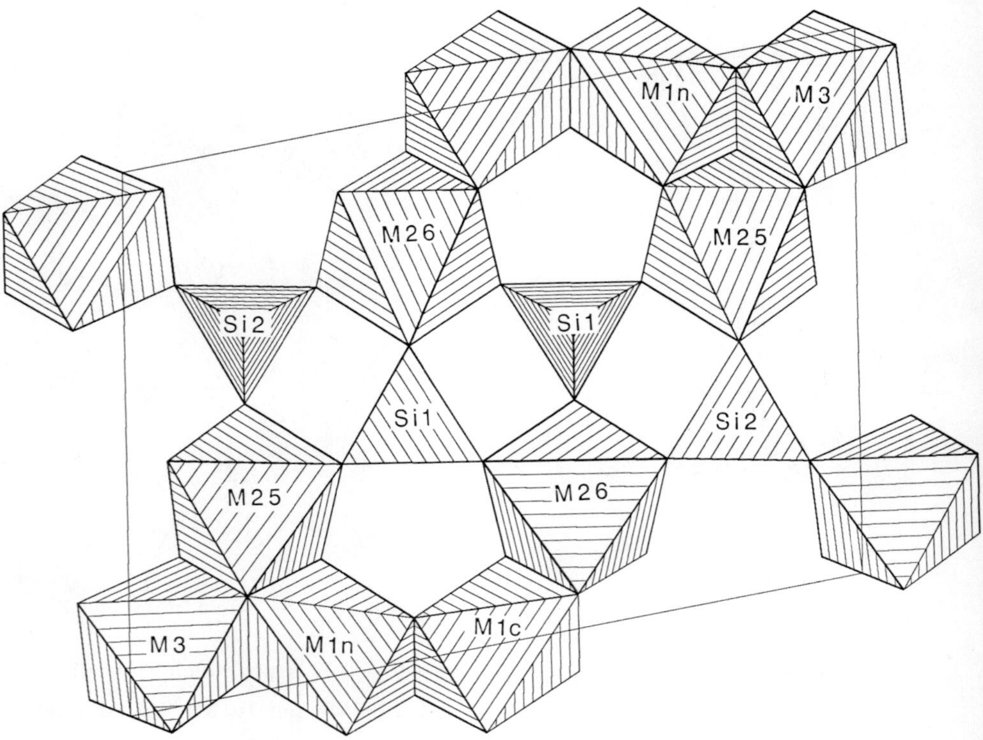

Figure 4.7d. Clinohumite, $(Mg,Fe)_9(SiO_4)_4(F,OH)_2$, *a*-axis projection, *b* vertical. There are five distinct octahedral sites all of which are in the general position with point symmetry 1 except $M1_c$ which lies on the inversion. There are also two distinct *Si* tetrahedra, both in the general position.

Table 4.7.2a. Humite Group Octahedral Sites.

End-Member	Norbergite		Chondrodite		
Site	M3	M2	M1	M2	M3
C.N.	6	6	6	6	6
Occupant	Mg	Mg	$Mg_{.95}Fe_{.05}$	Mg	Mg
Point Sym.	1	*m*	$\bar{1}$	1	1
Wyckoff Not.	8d	4c	2d	4e	4e
Frac.Coord.					
x	.9890	.9924	1/2	.0091	.4915
y	.6330	.9077	0	.1731	.8867
z	.4305	1/4	1/2	.3055	.0791
Distances					
Mean	2.068	2.104	2.170	2.116	2.078
σ	.075	.100	.014	.081	.072
Poly.Vol.	11.515	12.029	11.965	12.245	11.665
Q.E.	1.0174	1.0236	1.0277	1.0220	1.0179
Ang.Var.	56.5	75.6	100.6	74.0	59.2
Site Energy	-1085.	-1086.	-1055.	-1119.	-1095.

Table 4.7.2b. Humite Group Octahedral Sites (continued).

End-Member	Humite				Clinohumite				
Site	M1	M2	M3	M4	$M1_c$	$M1_n$	$M2_5$	$M2_6$	M3
C.N.	6	6	6	6	6	6	6	6	6
Occupant	$Mg_{.9}Fe_{.1}$	$Mg_{.9}Fe_{.1}$	$Mg_{.96}Fe_{.04}$	$Mg_{.99}Fe_{.01}$	$Mg_{.94}Fe_{.06}$	$Mg_{.91}Fe_{.09}$	$Mg_{.94}Fe_{.06}$	$Mg_{.92}Fe_{.08}$	$Mg_{.97}Fe_{.03}$
Point Sym.	1	*m*	1	1	$\bar{1}$	1	1	1	1
Wyckoff Not.	8d	4c	8d	8d	2d	4e	4e	4e	4e
Frac.Coord.									
x	.0017	.5108	.0087	.4925	1/2	.4977	.0101	.5101	.4939
y	.3773	.1540	.0976	.8665	0	.9463	.1398	.2503	.8780
z	.1767	1/4	.1092	.0278	1/2	.2738	.1703	.3888	.0428
Distances									
Mean	2.108	2.137	2.122	2.086	2.107	2.109	2.119	2.136	2.080
σ	.023	.086	.082	.074	.027	.024	.083	.083	.072
Poly.Vol.	11.970	12.483	12.345	11.780	11.925	11.978	12.280	12.483	11.703
Q.E.	1.0293	1.0291	1.0223	1.0189	1.0301	1.0297	1.0230	1.0283	1.0181
Ang.Var.	105.0	99.4	74.9	62.5	107.4	106.5	77.1	97.1	59.5
Site Energy	-1057.	-1135.	-1122.	-1099.	-1046.	-1058.	-1135.	-1131.	-1119.

Table 4.7.3. Humite Group Tetrahedral Sites.

Member	Norbergite	Chondr.	Humite		Clinohumite	
Site	Si	Si	Si1	Si2	Si1	Si2
C.N.	4	4	4	4	4	4
Occupant	Si	Si	Si	Si	Si	Si
Point Sym.	*m*	1	*m*	1	1	1
Wyckoff Not.	4c	4e	4c	8d	4c	4c
Frac.Coord.						
x	.4195	.0768	.0752	.5765	.0741	.0759
y	.7196	.1441	.9691	.2819	.0663	.1771
z	1/4	.7038	1/4	.1059	.3891	.8354
Distances						
Mean	1.630	1.633	1.629	1.627	1.626	1.638
σ	.012	.012	.009	.011	.004	.014
Poly.Vol.	2.193	2.202	2.188	2.180	2.175	2.219
Q.E.	1.0093	1.0102	1.0090	1.0096	1.0094	1.0109
Ang.Var.	41.26	45.17	38.18	42.47	39.65	48.56
Site Energy	-4321.	-4305.	-4293.	-4346.	-4299.	-4333.

Table 4.8.1. Titanite and Malayaite Unit Cells.

End-Member	Titanite	Malayaite
Formula	$CaTiSiO_5$	$CaSnSiO_5$
Form.Wt.	196.063	266.853
Density	3.517	4.546
Mol Volume	55.748	58.704
Z	4	4
Cryst.Sys.	Monoclinic	Monoclinic
Cryst.Class	$2/m$	$2/m$
Space Group	$P2_1/a$	$A2/a$
Cell Parameters		
a	7.069	7.149
b	8.722	8.906
c	6.566	6.667
β	113.86	113.3
Vol.	370.23	389.86
Ref.	Speer & Gibbs (1976)	Higgins & Ribbe (1977)

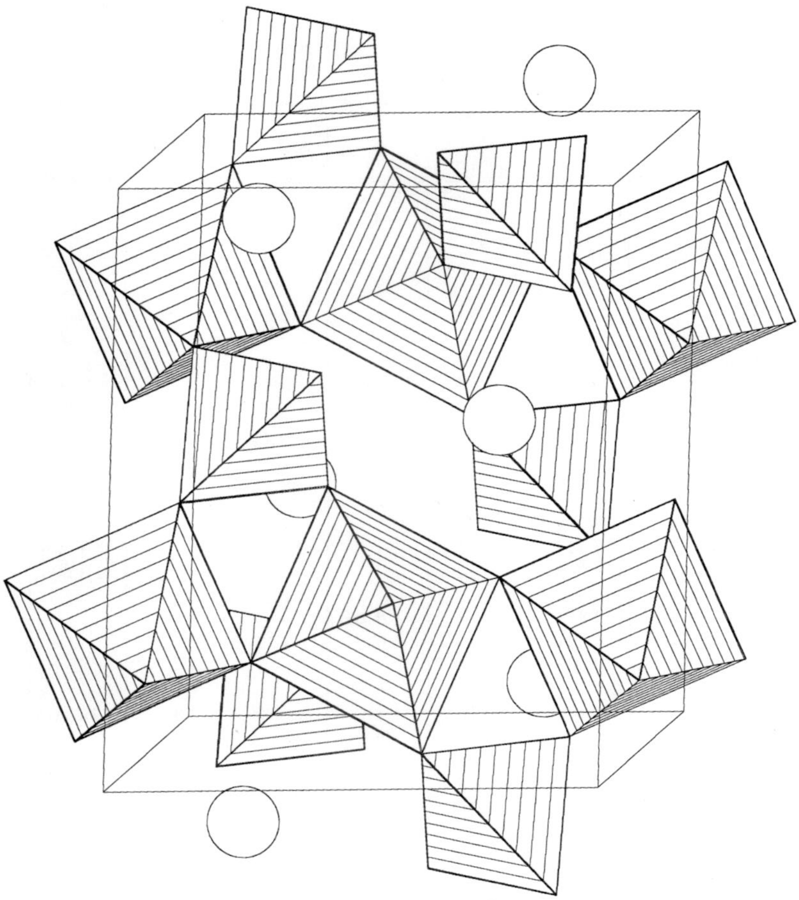

Figure 4.8. Titanite (sphene), $CaTiSiO_5$, *c*-axis perspective projection, *b* vertical. *Ca* (circle) is in seven-coordination, *Ti* is octahedral, and *Si* is tetrahedral. All cations are in the general position with point symmetry 1 in titanite, but in the isostructural malayaite ($CaSnSiO_5$) which has a higher symmetry, all cations are in special positions.

Table 4.8.2. Titanite and Malayaite Cation Sites.

End-Member	Titanite			Malayaite		
C.N.	7	6	4	7	6	4
Occupant	Ca	Ti	Si	Ca	Sn	Si
Point Sym.	1	1	1	2	$\bar{1}$	2
Wyckoff Not.	4e	4e	4e	4e	4c	4e
Frac.Coord.						
x	.2424	.5134	.7486	1/4	1/2	3/4
y	.9184	.7542	.9330	.9117	3/4	.9335
z	.7512	.2495	.7490	3/4	1/4	3/4
Distances						
O1		1.766			(2)1.947	
O1	2.281	1.974		2.241		
O2	2.428	1.991	1.646	(2)2.407	(2)2.088	(2)1.649
O3	2.399	2.014	1.647	(2)2.442	(2)2.092	(2)1.633
O3	2.672			(2)2.746		
O4	2.413	1.984	1.640			
O5	2.424	2.025	1.646			
O5	2.587					
Mean	2.458	1.959	1.645	2.490	2.042	1.641
σ	.130	.096	.003	.188	.074	.009
Poly.Vol.	19.713	9.978	2.273	20.688	11.300	2.253
Q.E.	-	1.0052	1.0032	-	1.0046	1.0044
Ang.Var.	-	7.56	12.12	-	8.45	17.36
Site Energy	-999.	-4160.	-4420.	-1007.	-3934.	-4448.

Table 4.9.1. Staurolite Unit Cell.

End-Member	Staurolite
Formula	$Fe_4Al_{18}Si_8O_{46}(OH)_2$
Form.Wt.	1703.73
Density	3.823
Mol Volume	445.67
Z	1
Cryst.Sys.	Monoclinic
Cryst.Class	$2/m$
Space Group	$C2/m$
Cell Parameters	
a	7.8713
b	16.6204
c	5.6560
β	90.0
Vol.	739.94
Ref.	Smith (1968)

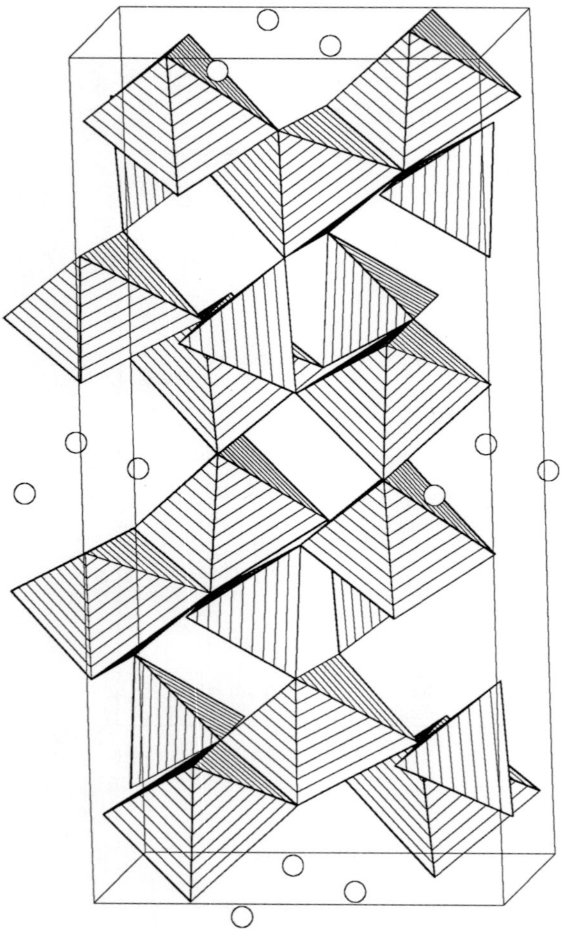

Figure 4.9. Staurolite, $Fe_4Al_{18}Si_8O_{46}(OH)_2$, c-axis perspective projection, b vertical. Fe (circle) and Si are in tetrahedral coordination. Al is in three distinct octahedral sites; Al1a and Al1b with point symmetry 2 and Al2 with point symmetry 1. Partially occupied sites U1, U2, Al3a, and Al3b are omitted for clarity.

Table 4.9.2. Staurolite Fully Occupied Cation Sites.

Site	Al1A	Al1B	Al2	Si
C.N.	6	6	6	4
Occupant	Al	Al	Al	Si
Point Sym.	2	2	1	1
Wyckoff Not.	4g	4h	8j	8j
Frac.Coord.				
x	1/2	1/2	.26356	.13414
y	.17511	.17477	.41042	.16612
z	0	1/2	.25122	.24902
Distances				
O1A			1.920	
O1B			1.921	
O2A	(2)1.938		1.930	1.634
O2B		(2)1.942	1.920	1.637
O3			1.873	1.651
O4	(2)1.900	(2)1.903		1.641
O5	(2)1.896	(2)1.897	1.867	
Mean	1.911	1.914	1.905	1.641
σ	.021	.022	.029	.008
Poly.Vol.	9.127	9.169	9.031	2.266
Q.E.	1.0133	1.0132	1.0139	1.0004
Ang.Var.	45.47	45.32	46.26	1.66
Site Energy	-2365.	-2365.	-2619.	-4361.
Model Charge	3.0	3.0	3.0	4.0

Table 4.9.3. Staurolite Partially Occupied Cation Sites.

Site	Fe	U(1)	U(2)	Al3A	Al3B
C.N.	4	6	6	6	6
Occupant	$Fe_{.64}Al_{.32}$	$Fe_{.68}Mn_{.32}$	$Fe_{.68}Mn_{.32}$	$Al_{.67}Fe_{.33}$	$Al_{.67}Fe_{.33}$
Occupancy	.916	.080	.038	.415	.282
Point Sym.	m	$2/m$	$2/m$	$2/m$	$2/m$
Wyckoff Not.	4i	2b	2d	2a	2c
Frac.Coord.					
x	.39281	1/2	1/2	0	0
y	0	0	0	0	0
z	.24815	0	1/2	0	1/2
Distances					
O1A	2.046	(2)2.114		(2)1.844	
O1B	2.043		(2)2.100		(2)1.855
O3				(4)2.037	(4)2.060
O5	(2)1.972	(4)2.191	(4)2.195		
Mean	2.008	2.165	2.163	1.972	1.992
σ	.042	.040	.049	.100	.106
Poly.Vol.	4.141	12.960	12.957	10.125	10.441
Q.E.	1.0026	1.0299	1.0286	1.0092	1.0083
Ang.Var.	11.5	90.4	85.8	16.6	12.2
Site Energy	-1395.	+12.	+6.	-622.	-369.
Model Charge	2.36	0.16	0.08	1.108	.753

Chapter 5

SORO-AND CYCLO-SILICATES

Table 5.1.1. Epidote Group Unit Cells.

End-Member	Zoisite	Clinozoisite	Epidote	Epidote	Allanite
Formula	$Ca_2Al_3Si_3O_{12}(OH)$	$Ca_2Al_3Si_3O_{12}(OH)$	$Ca_2Al_{2.16}Fe_{.84}Si_3O_{12}(OH)$	$Ca_2Al_{2.60}Fe_{.40}Si_3O_{12}(OH)$	$Ca_{1.26}RE_{.74}Al_{1.83}Fe_{1.17}Si_3O_{12}(OH)$
Form.Wt.	454.363	454.363	478.610	465.909	565.2
Density	3.336	3.321	3.465	3.392	3.960
Mol Volume	136.19	136.83	138.146	137.370	142.737
Z	4	2	2	2	2
Cryst.Sys.	Orthorhombic	Monoclinic	Monoclinic	Monoclinic	Monoclinic
Cryst.Class	*mmm*	$2/m$	$2/m$	$2/m$	$2/m$
Space Group	*Pnma*	$P2_1/m$	$P2_1/m$	$P2_1/m$	$P2_1/m$
Cell Parameters					
a	16.212	8.879	8.8877	8.8802	8.927
b	5.559	5.583	5.6275	5.6043	5.761
c	10.036	10.155	10.1517	10.1511	10.150
β	90.	115.50	115.383	115.455	114.77
Vol.	904.47	454.36	458.73	456.15	473.97
Ref.	Dollase (1968)	Dollase (1968)	Gabe et al. (1973)	Gabe et al. (1973)	Dollase (1971)

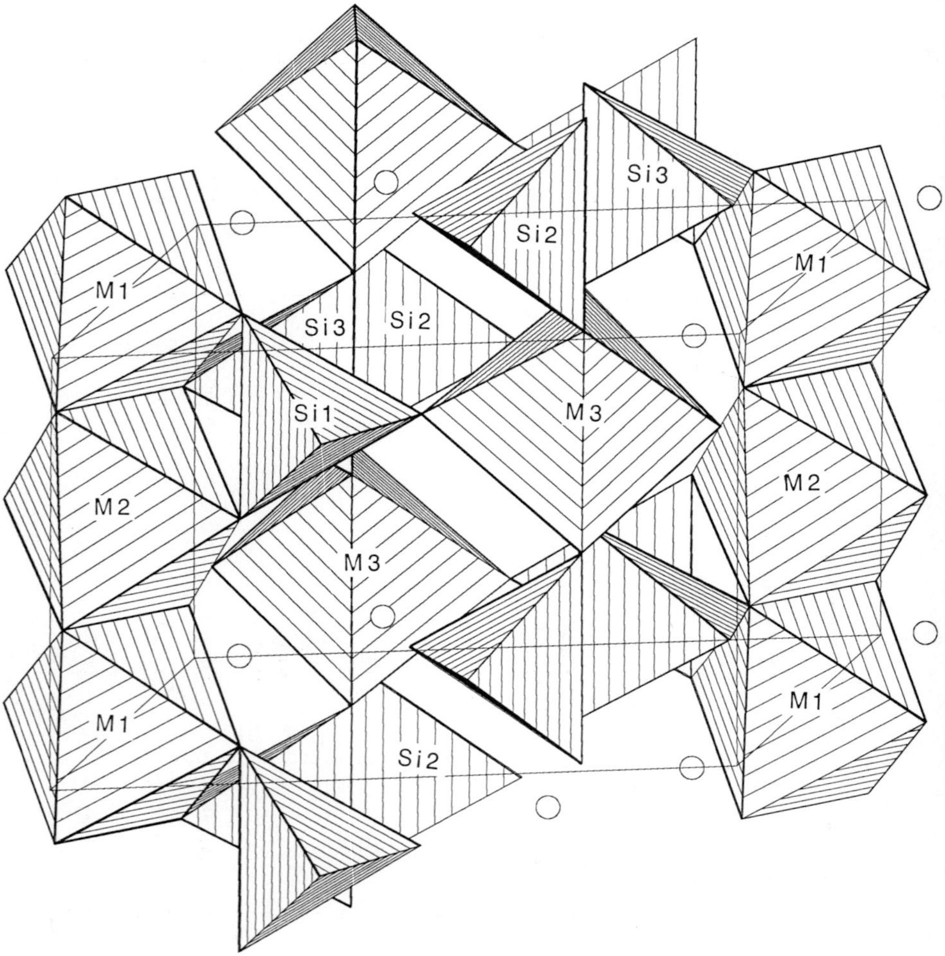

Figure 5.1. Epidote, $Ca_2(Al,Fe)_3Si_3O_{12}(OH)$, perspective *c*-axis projection, *b* vertical. There are two distinct *Ca* sites, A1 and A2 (circles), both with eight-coordination and point symmetry *m*. The rare earth elements occupy the A2 site in allanite. The unusually large electrostatic energy of this site for its size may explain why this site strongly fractionates the lanthanides in natural systems. The three octahedral sites M1, M2, and M3, are occupied by *Al* and lesser amounts of Fe^{+3}. M1 and M2 lie on inversion centers, and M3 lies on a mirror. There are three distinct *Si* tetrahedra, T1, T2, and T3, all of which lie on the mirror.

Table 5.1.2. Epidote Group Ca Sites.

End-Member	Zoisite		Clinozoisite		Epidote		Epidote		Allanite	
Site	A1	A2	A1	A2	A1	A2	A1	A2	A1	A2
C.N.	9	7	9	8	9	8	9	8	9	11
Occupant	Ca	Ca	Ca	Ca	Ca	Ca	Ca	Ca	Ca	$RE_{.74}Ca_{.26}$
Point Sym.	*m*	*m*	*m*	*m*	*m*	*m*	*m*	*m*	*m*	*m*
Wyckoff Not.	4c	4c	2e	2e	2e	2e	2e	2e	2e	2e
Frac.Coord.										
x	.3667	.4518	.7617	.6063	.75715	.60486	.75972	.60660	.7585	.5936
y	1/4	1/4	3/4	3/4	3/4	3/4	3/4	3/4	3/4	3/4
z	.4376	.1150	.1555	.4234	.15156	.42399	.15373	.42363	.1517	.4286
Distances										
O1(2)	2.504		2.491		2.459		2.478		2.373	
O2(2)		2.789		2.818		2.784		2.810		2.642
O2(2)		2.521		2.543		2.527		2.536		2.516
O3(2)	2.416	2.468	2.368	2.532	2.323	2.653	2.345	2.575	2.337	2.801
O5(1)	2.588		2.522		2.556		2.534		2.592	
O6(1)	2.552		2.745		2.861		2.789		2.911	
O7(1)	2.251	2.305	2.283	2.267	2.295	2.248	2.284	2.262	2.369	2.329
O8(2)										3.017
O8(2)										3.127
O9(2)	2.916		2.952		3.000		2.973		3.112	
O10(1)				2.575		2.531		2.551		2.611
Mean	2.562	2.551	2.575	2.579	2.586	2.588	2.578	2.582	2.613	2.729
σ	.223	.178	.250	.177	.291	.174	.268	.174	.338	.248
Poly.Vol.	28.095	31.011	27.608	32.890	27.249	33.079	27.405	32.977	27.531	54.860
Site Energy	-957.	-1040.	-958.	-1053.	-961.	-1071.	-972.	-1044.	-955.	-1105.
Model Charge	2.0	2.0	2.0	2.0	2.0	2.0	2.0	2.0	2.0	2.0

Table 5.1.3. Epidote Group Octahedral Sites.

End-Member	Zoisite		Clinozoisite			Epidote			Epidote			Allanite		
Site	M1,2	M3	M1	M2	M3	M1	M2	M3	M1	M2	M3	M1	M2	M3
C.N.	6	6	6	6	6	6	6	6	6	6	6	6	6	6
Occupant	Al	Al	Al	Al	Al,Fe	Al	Al	Al,Fe	Al	Al	Al,Fe	Al,Fe	Al	Fe,Al
Point Sym.	1	*m*	$\bar{1}$	$\bar{1}$	*m*	$\bar{1}$	$\bar{1}$	*m*	$\bar{1}$	$\bar{1}$	*m*	$\bar{1}$	$\bar{1}$	*m*
Wyckoff Not.	8d	4c	2a	2c	2e	2a	2c	2e	2a	2c	2e	2a	2c	2e
Frac.Coord.														
x	.2496	.1054	0	0	.2873	0	0	.29386	0	0	.29085	0	0	.3030
y	.9971	3/4	0	0	1/4	0	0	1/4	0	0	1/4	0	0	1/4
z	.1899	.3006	0	1/2	.2238	0	1/2	.22419	0	1/2	.22415	0	1/2	.2148
Distances														
O1	(1)1.964	(2)2.133	(2)1.930		(2)2.184	(2)1.939		(2)2.224	(2)1.931		(2)2.200	(2)1.992		(2)2.304
O2(2)		1.965			1.926			(2)1.985			(2)1.956			(2)2.195
O3	(1)1.850			(2)1.859			(2)1.854			(2)1.858			(2)1.876	
O4	(1)1.843	(1)1.822	(2)1.850		(1)1.861	(2)1.843		(1)1.935	(2)1.847		(1)1.903	(2)1.878		(1)2.002
O5	(1)1.900		(2)1.936			(2)1.956			(2)1.943			(2)2.026		
O6	(1)1.926			(2)1.923			(2)1.927			(2)1.926			(2)1.920	
O8		(1)1.784			(1)1.781			(1)1.860			(1)1.810			(1)1.941
O10	(1)1.849			(2)1.852			(2)1.870			(2)1.864			(2)1.914	
Mean	1.888	1.967	1.906	1.878	1.977	1.913	1.883	2.036	1.907	1.883	2.004	1.965	1.904	2.157
σ	.050	.148	.043	.035	.169	.054	.034	.153	.047	.034	.161	.069	.021	.153
Poly.Vol.	8.899	9.866	9.146	8.773	10.009	9.252	8.853	10.864	9.167	8.847	10.395	10.053	9.110	12.637
Q.E.	1.0066	1.0237	1.0065	1.0045	1.0259	1.0065	1.0045	1.0283	1.0064	1.0042	1.0271	1.0055	1.0064	1.0433
Ang.Var.	20.1	54.9	19.7	14.0	58.1	17.9	14.0	74.8	18.9	13.1	66.3	11.9	22.1	125.5
Site Energy	-2573.	-2541.	-2435.	-2633.	-2545.	-2410.	-2614.	-2413.	-2427.	-2684.	-2452.	-2298.	-2640.	-2203.
Model Charge	3.0	3.0	3.0	3.0	3.0	3.0	3.0	3.0	3.0	3.0	3.0	3.0	3.0	3.0

Table 5.1.4. Epidote Group Tetrahedral Sites.

End-Member	Zoisite			Clinozoisite			Epidote			Epidote			Allanite		
Site	T1	T2	T3	T1	T2	T3	T1	T2	T3	T1	T2	T3	T1	T2	T3
C.N.	4	4	4	4	4	4	4	4	4	4	4	4	4	4	4
Occupant	Si	Si	Si	Si	Si	Si	Si	Si	Si	Si	Si	Si	Si	Si	Si
Point Sym.	*m*	*m*	*m*	*m*	*m*	*m*	*m*	*m*	*m*	*m*	*m*	*m*	*m*	*m*	*m*
Wyckoff Not.	4c	4c	4c	2e	2e	2e	2e	2e	2e	2e	2e	2e	2e	2e	2e
Frac.Coord.															
x	.0816	.4104	.1601	.3382	.6776	.1822	.33959	.68429	.18393	.33880	.68054	.18304	.3389	.6866	.1880
y	1/4	3/4	1/4	3/4	1/4	3/4	3/4	1/4	3/4	3/4	1/4	3/4	3/4	1/4	3/4
z	.1064	.2821	.4356	.0478	.2753	.3158	.04766	.27454	.31837	.04804	.27507	.31686	.0369	.2799	.3240
Distances															
O1(2)	1.656			1.652			1.650			1.652			1.645		
O2(2)			1.619			1.629			1.627			1.625			1.604
O3(2)		1.621			1.620			1.618			1.619			1.631	
O5(1)			1.651			1.662			1.668			1.667			1.657
O6(1)			1.672			1.657			1.638			1.648			1.661
O7(1)	1.586			1.566			1.564			1.562			1.581		
O8(1)		1.580			1.593			1.588			1.592			1.603	
O9(1)	1.640	1.624		1.627	1.627		1.634	1.631		1.633	1.627		1.651	1.627	
Mean	1.635	1.612	1.640	1.625	1.615	1.644	1.624	1.614	1.640	1.625	1.614	1.641	1.630	1.623	1.632
σ	.033	.021	.026	.041	.015	.018	.041	.019	.019	.043	.016	.020	.033	.014	.032
Poly.Vol.	2.226	2.143	2.227	2.190	2.157	2.261	2.190	2.155	2.249	2.191	2.155	2.251	2.217	2.192	2.201
Q.E.	1.0049	1.0018	1.0114	1.0034	1.0014	1.0060	1.0033	1.0009	1.0046	1.0035	1.0012	1.0055	1.0027	1.0011	1.0086
Ang.Var.	17.5	7.4	44.3	10.5	6.0	23.9	9.9	3.7	18.4	10.5	5.0	22.2	8.9	4.1	33.5
Site Energy	-4368.	-4529.	-4400.	-4380.	-4540.	-4395.	-4348.	-4542.	-4427.	-4373.	-4564.	-4390.	-4286.	-4504.	-4518.
Model Charge	4.0	4.0	4.0	4.0	4.0	4.0	4.0	4.0	4.0	4.0	4.0	4.0	4.0	4.0	4.0

Table 5.2.1. Melilite Group Unit Cells.

End-Member	Na-Melilite	Gehlenite	Akermanite
Formula	$CaNaAlSi_2O_7$	$Ca_2Al(Al,Si)_2O_7$	$Ca_2MgSi_2O_7$
Form.Wt.	258.219	274.205	272.640
Density	2.912	3.006	2.944
Mol Volume	88.662	91.220	92.619
Z	2	2	2
Cryst.Sys.	Tetragonal	Tetragonal	Tetragonal
Cryst.Class	$\bar{4}2m$	$\bar{4}2m$	$\bar{4}2m$
Space Group	$P\bar{4}2_1m$	$P\bar{4}2_1m$	$P\bar{4}2_1m$
Cell Parameters			
a	7.6344	7.7173	7.835
c	5.0513	5.0860	5.010
Vol.	294.41	302.91	307.55
Ref.	Louisnathan (1970b)	Louisnathan (1970a)	Kimata & Ii (1981)

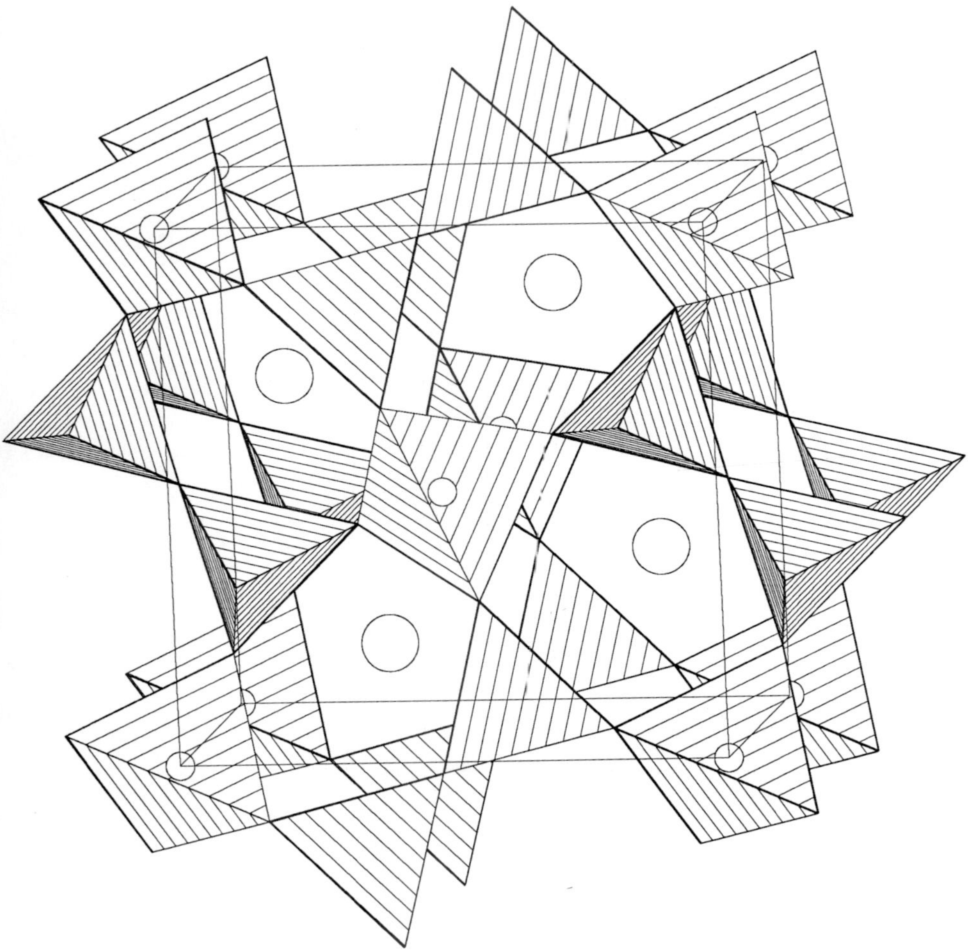

Figure 5.2. Melilite, $CaNaAlSi_2O_7$, perspective c-axis projection. The $Ca-Na$ site is eight-coordinated with point symmetry m. There are two distinct tetrahedral sites. The larger one (circle) is occupied by Al (or Mg in akermanite) and lies at the origin with point symmetry $\bar{4}$. The smaller one is occupied by Si (or Al and Si in gehlenite) and has point symmetry m to form part of an Si_2O_7 group. There are no octahedral sites, and all oxygens are shared between two tetrahedral sites, so the structure might also be considered a framework silicate.

Table 5.2.2. Melilite Ca-Na Sites.

End-Member	Na-Melilite	Gehlenite	Akermanite
C.N.	8	8	8
Occupant	$Ca_{.5}Na_{.5}$	Ca	Ca
Point Sym.	*m*	*m*	*m*
Wyckoff Not.	4e	4e	4e
Frac.Coord.			
x	.3399	.3375	.3318
y	.1601	.1625	.1682
z	.5134	.5110	.5067
Distances			
O1(1)	2.453	2.416	2.484
O2(1)	2.470	2.430	2.465
O2(2)	2.572	2.576	2.712
O3(2)	2.463	2.438	2.425
O3(2)	2.793	2.816	2.695
Mean	2.572	2.563	2.577
σ	.144	.168	.137
Poly.Vol.	32.893	32.475	32.909
Site Energy	-604.	-991.	-952.
Model Charge	1.5	2.0	2.0

Table 5.2.3. Melilite Al-Mg Tetrahedral Sites.

End-Member	Na-Melilite	Gehlenite	Akermanite
C.N.	4	4	4
Occupant	Al^{+3}	Al^{+3}	Mg^{+2}
Point Sym.	$\bar{4}$	$\bar{4}$	$\bar{4}$
Wyckoff Not.	2a	2a	2a
Frac.Coord.			
x	0	0	0
y	0	0	0
z	0	0	0
Distances			
O3(4)	1.762	1.785	1.915
σ	0	0	0
Poly.Vol.	2.788	2.916	3.599
Q.E.	1.0040	1.0010	1.0010
Ang.Var.	16.8	3.6	5.8
Site Energy	-2442.	-2431.	-1184.
Model Charge	3.0	3.0	2.0

Table 5.2.4. Melilite Al-Si Tetrahedral Sites.

End-Member	Na-Melilite	Gehlenite	Akermanite
C.N.	4	4	4
Occupant	Si	$Al_{.5}Si_{.5}$	Si
Point Sym.	*m*	*m*	*m*
Wyckoff Not.	4e	4e	4e
Frac.Coord.			
x	.1416	.1431	.1397
y	.3584	.3569	.3603
z	.9531	.9528	.9352
Distances			
O1(1)	1.648	1.719	1.650
O2(1)	1.577	1.680	1.595
O3(2)	1.631	1.683	1.616
Mean	1.622	1.691	1.619
σ	.031	.019	.023
Poly.Vol.	2.163	2.431	2.145
Q.E.	1.0081	1.0143	1.0110
Ang.Var.	36.6	61.5	47.1
Site Energy	-4341.	-3338.	-4378.
Model Charge	4.0	3.5	4.0

Table 5.3.1. Beta Orthosilicate Unit Cells.

End-Member	Mg_2SiO_4	Co_2SiO_4
Form.Wt.	140.708	209.950
Density	3.4729	5.044
Mol Volume	40.515	41.628
Z	8	8
Cryst.Sys.	Orthorhombic	Orthorhombic
Cryst.Class	*mmm*	*mmm*
Space Group	*Imma*	*Imma*
Cell Parameters		
a	5.6983	5.753
b	11.4380	11.524
c	8.2566	8.340
Vol.	538.14	552.92
Ref.	Horiuchi & Sawamoto (1981)	Morimoto et al. (1974)

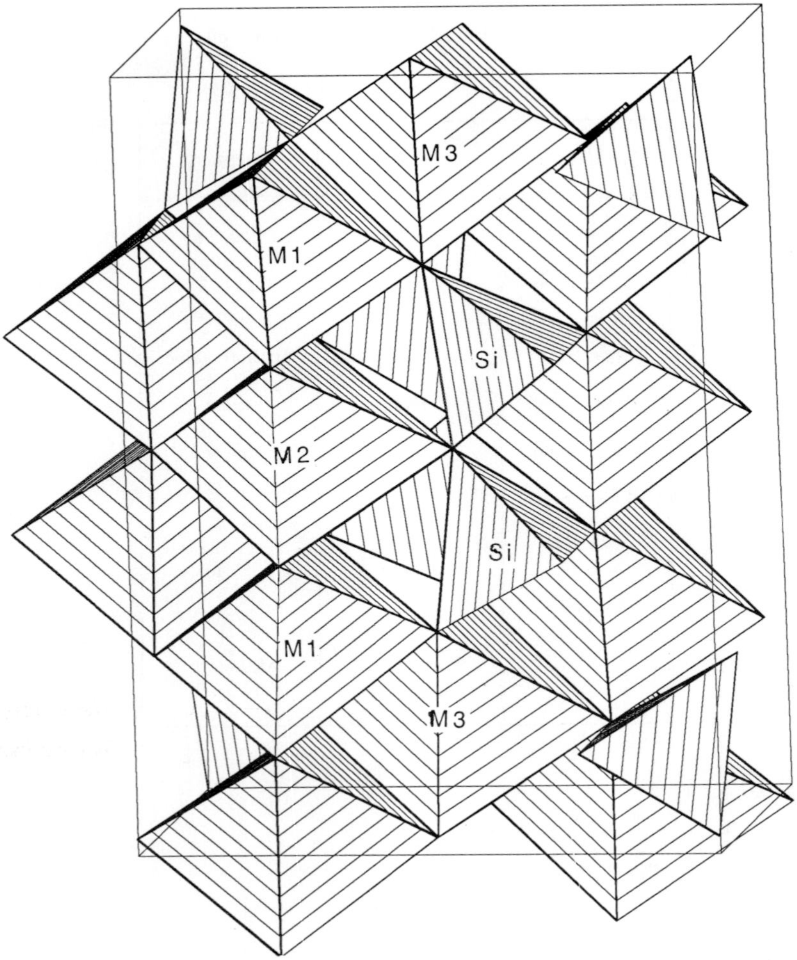

Figure 5.3. Beta Orthosilicate, Mg_2SiO_4, perspective a-axis projection, c horizontal. This structure is sometimes referred to as "β spinel" or "modified spinel." It is not a spinel, nor even an orthosilicate. This is a sorosilicate with a clear, isolated Si_2O_7 group, both Si atoms being equivalent with point symmetry m. This site is exceptionally distorted for an Si tetrahedron. There are three distinct Mg octahedra, M1 with point symmetry $2/m$, M2 with point symmetry mm, and M3 with point symmetry 2. One of the oxygens, O1, is not bonded to Si and may be a potential site for hydroxyl in the mantle.

Table 5.3.2. Beta Orthosilicate M Sites.

End-Member	Mg_2SiO_4			Co_2SiO_4		
Site	M1	M2	M3	M1	M2	M3
C.N.	6	6	6	6	6	6
Occupant	Mg	Mg	Mg	Co	Co	Co
Point Sym.	$2/m$	mm	2	$2/m$	mm	2
Wyckoff Not.	4c	4e	8f	4c	4e	8f
Frac.Coord.						
x	0	0	1/4	0	0	1/4
y	0	1/4	.1276	0	1/4	.1241
z	0	.9701	1/4	0	-.0286	1/4
Distances						
O1(1)		2.035			2.052	
O1(2)			2.016			2.061
O2(1)		2.095			2.135	
O3(2)	2.115		2.123	2.147		2.147
O4(4)	2.046	2.093		2.086	2.128	
O4(2)			2.128			2.156
Mean	2.069	2.084	2.089	2.106	2.117	2.121
σ	.036	.024	.056	.032	.031	.047
Poly.Vol.	11.731	11.966	12.039	12.321	12.549	12.614
Q.E.	1.0050	1.0055	1.0072	1.0077	1.0051	1.0064
Ang.Var.	15.2	19.3	23.4	24.5	17.4	21.0
Site Energy	-1172.	-1165.	-1193.	-1132.	-1141.	-1164.

Table 5.3.3. Beta Orthosilicate Tetrahedral Sites.

End-Member	Mg_2SiO_4	Co_2SiO_4
C.N.	4	4
Occupant	Si	Si
Point Sym.	*m*	*m*
Wyckoff Not.	8i	8i
Frac.Coord.		
x	0	.1331
y	.1198	.1211
z	.6168	1/4
Distances		
O2(1)	1.701	1.697
O3(1)	1.638	1.622
O4(2)	1.632	1.621
Mean	1.651	1.641
σ	.034	.038
Poly.Vol.	2.297	2.257
Q.E.	1.0037	1.0029
Ang.Var.	14.7	11.6
Site Energy	-4322.	-4366.

Table 5.4.1. Lawsonite Unit Cell.

End-Member	Lawsonite
Formula	$CaAl_2Si_2O_7(OH)_2H_2O$
Form.Wt.	314.241
Density	3.088
Mol Volume	101.76
Z	4
Cryst.Sys.	Orthorhombic
Cryst.Class	*mmm*
Space Group	*Ccmm*
Cell Parameters	
a	8.795
b	5.847
c	13.142
Vol.	675.82
Ref.	Baur (1978)

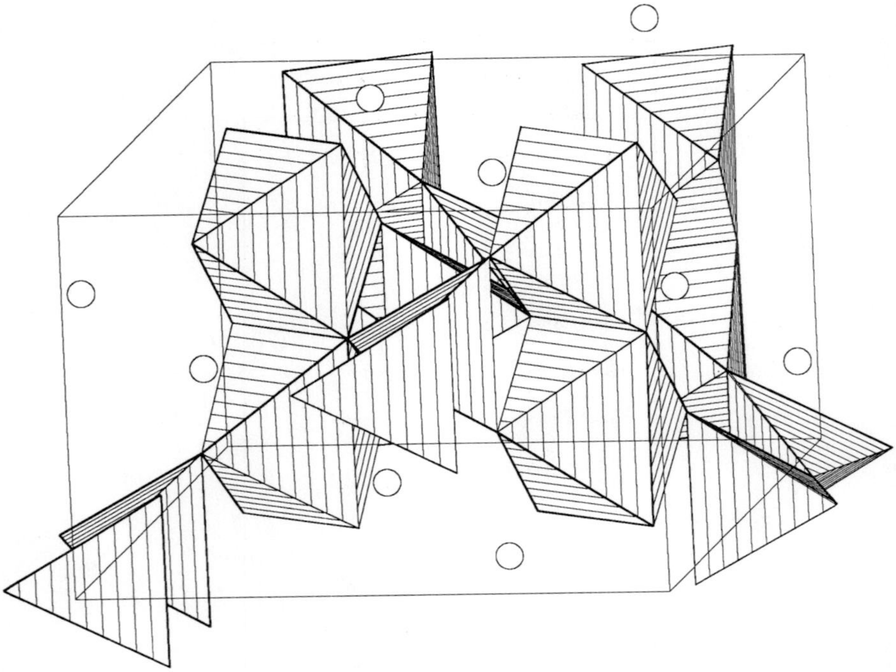

Figure 5.4. Lawsonite, $CaAl_2Si_2O_7(OH)_2H_2O$, perspective c-axis projection, b vertical. Ca (circle) is in distorted six-coordination with point symmetry mm. Al is in octahedral coordination with point symmetry $\bar{1}$, and Si has point symmetry m in tetrahedral coordination as part of an Si_2O_7 group.

Table 5.4.2. Lawsonite Cation Sites.

End-Member	Lawsonite		
Site	Ca	Al	Si
C.N.	6	6	4
Occupant	Ca	Al	Si
Point Sym.	*mm*	$\bar{1}$	*m*
Wyckoff Not.	4c	8d	8f
Frac.Coord.			
x	.33305	1/4	.9804
y	0	1/4	0
z	1/4	0	.13298
Distances			
Mean	2.421	1.913	1.633
σ	.038	.042	.020
Poly.Vol.	18.315	9.190	2.219
Q.E.	1.0225	1.0112	1.0049
Ang.Var.	78.39	36.74	19.61
Site Energy	-1170.	-2679.	-4779.
Model Charge	2.0	3.0	4.0

Table 5.5.1. Tourmaline Group Unit Cells.

End-Member	Dravite	Schorl	Elbaite
Formula	$(Na_{.44}Ca_{.36}Mg_{.18})$ $(Mg_{1.87}V_{.76}Cr_{.19}Fe_{.18})$ $(Al_{5.56}V_{.38})B_3$ $(Si_{5.63}Al_{.37})O_{27}$ $(O_{1.10}OH_{2.58}F_{.32})$	$(Na_{.88}Ca_{.11})$ $(Fe_{2.06}Al_{.44}Mg_{.11}$ $Ca_{.07}Ti_{.07}Li_{.07})$ $(Al_{5.61}Fe_{.39})B_3$ Si_6O_{27} $(O,OH,F)_4$	$(Na_{.56}Ca_{.14}Mn_{.15})$ $(Al_{1.59}Li_{1.25}Mn_{.13})$ Al_6B_3 Si_6O_{27} $(O_{.52}OH_{2.87}F_{.60})$
Form.Wt.	1001.6	1043.3	948.8
Density	3.142	3.263	3.063
Mol Volume	318.76	319.71	309.80
Z	3	3	3
Cryst.Sys.	Trigonal	Trigonal	Trigonal
Cryst.Class	$3m$	$3m$	$3m$
Space Group	$R3m$	$R3m$	$R3m$
Cell Parameters			
a	15.967	15.992	15.838
c	7.191	7.190	7.1032
Vol.	1587.7	1592.5	1543.1
Ref.	Foit & Rosenberg (1979)	Fortier & Donnay (1975)	Donnay & Barton (1972)

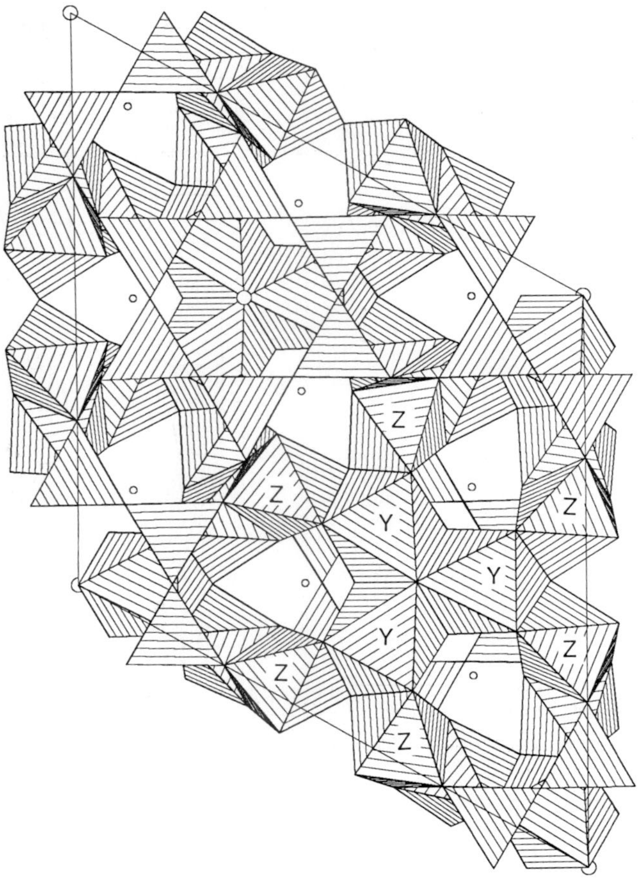

Figure 5.5. Tourmaline, $(Na, Ca)(Mg, Fe)_3Al_6B_3Si_6O_{27}(OH)_4$, c-axis projection. The X site (large circle) is nine-coordinated with point symmetry $3m$ and is usually occupied by Na or Ca. The Y site is octahedral with point symmetry m and is usually occupied by Mg or Fe. The Z site is also octahedral with point symmetry 1 and is usually occupied by Al. The B site (small circle) is planar three-coordinated with point symmetry m. The Si site is tetrahedral with point symmetry 1 and forms part of an Si_6O_{18} ring.

Table 5.5.2. Tourmaline Group X Sites.

End-Member	Dravite	Schorl	Elbaite
C.N.	9	9	9
Occupant	$Na_{.44}Ca_{.36}Mg_{.18}$	$Na_{.88}Ca_{.11}$	$Na_{.56}Ca_{.14}Mn_{.15}$
Point Sym.	$3m$	$3m$	$3m$
Wyckoff Not.	3a	3a	3a
Frac.Coord.			
x	0	0	0
y	0	0	0
z	.21764	.22353	.2347
Distances			
O2(3)	2.536	2.527	2.458
O4(3)	2.780	2.800	2.816
O5(3)	2.704	2.744	2.738
Mean	2.673	2.690	2.671
σ	.108	.125	.163
Poly.Vol.	31.56	32.32	31.46
Site Energy	-600.	-349.	-379.
Model Charge	1.54	1.11	1.14

Table 5.5.3. Tourmaline Group Y Sites.

End-Member	Dravite	Schorl	Elbaite
C.N.	6	6	6
Occupant	$Mg_{.62}V_{.25}Cr_{.06}Fe_{.06}$	$Fe_{.68}Al_{.15}Mg_{.04}$ $Ca_{.02}Li_{.02}Ti_{.02}$	$Al_{.53}Li_{.42}Mn_{.04}$
Point Sym.	*m*	*m*	*m*
Wyckoff Not.	9b	9b	9b
Frac.Coord.			
x	.12353	.12566	.1234
y	x/2	x/2	x/2
z	.63364	.62792	.6348
Distances			
O1(1)	1.971	2.069	1.984
O2(2)	2.018	2.013	1.985
O3(1)	2.004	2.175	2.173
O6(2)	2.137	2.046	1.984
Mean	2.025	2.060	2.016
σ	.057	.060	.077
Poly.Vol.	10.715	11.247	10.524
Q.E.	1.0230	1.0252	1.0262
Ang.Var.	75.6	80.1	81.2
Site Energy	-1609.	-1387.	-1236.
Model Charge	2.32	2.33	2.11

Table 5.5.4. Tourmaline Group Z Sites.

End-Member	Dravite	Schorl	Elbaite
C.N.	6	6	6
Occupant	$Al_{.93}V_{.07}$	$Al_{.93}Fe_{.07}$	Al
Point Sym.	1	1	1
Wyckoff Not.	18c	18c	18c
Frac.Coord.			
x	.29782	.29883	.2964
y	.26152	.26171	.2598
z	.61011	.61158	.6105
Distances			
O3(1)	1.995	1.973	1.955
O6(1)	1.893	1.878	1.838
O7(1)	1.900	1.890	1.894
O7(1)	1.955	1.969	1.946
O8(1)	1.932	1.929	1.900
O8(1)	1.900	1.890	1.896
Mean	1.929	1.922	1.905
σ	.040	.042	.042
Poly.Vol.	9.379	9.297	9.031
Q.E.	1.0141	1.0127	1.0140
Ang.Var.	47.9	42.5	46.9
Site Energy	-2493.	-2397.	-2503.
Model Charge	3.0	2.93	3.0

Table 5.5.5. Tourmaline Group B Sites.

End-Member	Dravite	Schorl	Elbaite
C.N.	3	3	3
Occupant	B	B	B
Point Sym.	*m*	*m*	*m*
Wyckoff Not.	9b	9b	9b
Frac.Coord.			
x	.10976	.11029	.1092
y	2x	2x	2x
z	.45178	.45461	.4548
Distances			
O2(1)	1.369	1.366	1.331
O8(2)	1.376	1.382	1.376
Mean	1.374	1.376	1.361
σ	.004	.009	.026
Site Energy	-3112.	-3097.	-3151.
Model Charge	3.0	3.0	3.0

Table 5.5.6. Tourmaline Group Si Sites.

End-Member	Dravite	Schorl	Elbaite
C.N.	4	4	4
Occupant	Si	Si	Si
Point Sym.	1	1	1
Wyckoff Not.	18c	18c	18c
Frac.Coord.			
x	.19173	.19177	.1917
y	.18990	.18986	.1896
z	0	0	0
Distances			
O4(1)	1.630	1.627	1.620
O5(1)	1.645	1.639	1.629
O6(1)	1.613	1.601	1.605
O7(1)	1.612	1.610	1.611
Mean	1.625	1.620	1.616
σ	.016	.017	.010
Poly.Vol.	2.195	2.175	2.161
Q.E.	1.0024	1.0019	1.0019
Ang.Var.	9.8	7.6	7.9
Site Energy	-4247.	-4425.	-4401.
Model Charge	3.94	4.0	4.0

Table 5.6.1. Vesuvianite Unit Cell.

End-Member	Vesuvianite
Formula	$Ca_{17.6}(Ca_{1.0}Fe_{1.0})Al_{4.0}$ $(Mg_{1.1}Al_{6.0}Ti_{0.2}Mn_{0.2}Fe_{0.8})$ $(Al_{0.2}Si_{17.8})(O_{68.5}(OH)_{6.3}F_{3.2})$
Form.Wt.	2934.7
Density	3.429
Mol Volume	427.9
Z	2
Cryst.Sys.	Tetragonal
Cryst.Class	$4mm$
Space Group	$P\,4/nnc$
Cell Parameters	
a	15.5333
c	11.7778
Vol.	2841.79
Ref.	Allen (1985)

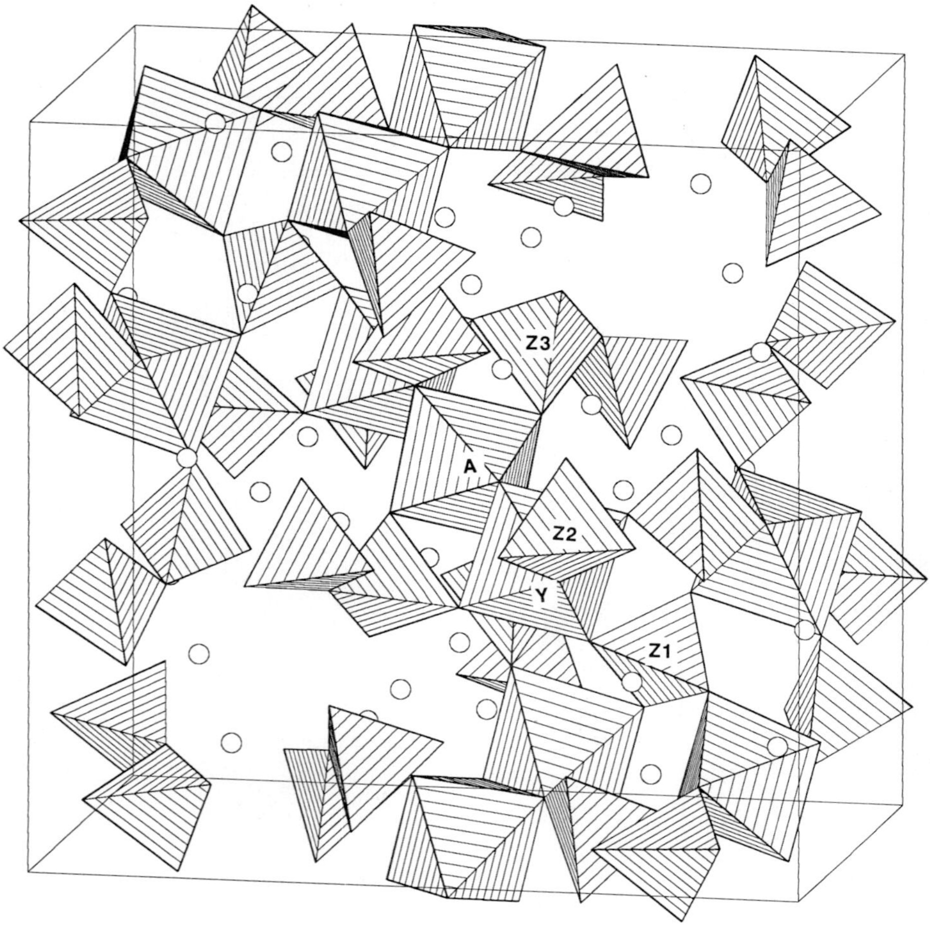

Figure 5.6. Vesuvianite (Idocrase), $Ca_{17.6}(CaFe)Al_4(Mg_{1.1}Al_{6.0}Ti_{0.2}Mn_{0.2}Fe_{0.8})(Al_{0.2}Si_{17.8})(O_{68.5}(OH)_{6.3}F_{3.2})$, c-axis perspective projection. There are three distinct Ca sites (circles), X1 with point symmetry 222, and X2 and X3, each with point symmetry 1. There are two distinct *Al* octahedra, the A site has point symmetry $\bar{1}$, and the Y site has point symmetry 1. There are three distinct *Si* tetrahedra, Z1 with point symmetry $\bar{4}$, and Z2 and Z3 both with point symmetry 1. The partially occupied B and C sites are omitted for clarity.

Table 5.6.2. Vesuvianite Ca Sites.

Site	X1	X2	X3
C.N.	8	7	8
Occupant	Ca	Ca	Ca
Point Sym.	222	1	1
Wyckoff Not.	4c	16k	16k
Frac.Coord.			
x	3/4	.8110	.8996
y	1/4	.0438	.8202
z	1/4	.3795	.8866
Distances			
Mean	2.421	2.402	2.499
σ	.105	.060	.080
Poly.Vol.	24.405	18.925	27.002
Site Energy	-754.	-860.	-1164.
Model Charge	2.0	2.0	2.0

Table 5.6.3. Vesuvianite Octahedral Sites.

Site	A	Y
C.N.	6	6
Occupant	Al	Al,Fe
Point Sym.	$\bar{1}$	1
Wyckoff Not.	8f	16k
Frac.Coord.		
x	0	.8872
y	0	.1214
z	0	.1264
Distances		
Mean	1.888	1.949
σ	.040	.061
Poly.Vol.	8.893	9.754
Q.E.	1.0068	1.0084
Ang.Var.	21.5	25.8
Site Energy	-2705.	-2407.
Model Charge	3.0	3.0

Table 5.6.4. Vesuvianite B and C Sites.

Site	B	C
C.N.	5	8
Occupant	Fe	Ca
Occupancy	0.5	0.5
Point Sym.	4	4
Wyckoff Not.	4e	4e
Frac.Coord.		
x	3/4	3/4
y	3/4	3/4
z	.0553	.1408
Distances		
Mean	2.107	2.474
σ	.062	.195
Poly.Vol.	6.604	32.133
Site Energy	-237.	-105.
Model Charge	1.0	1.0

Table 5.6.5. Vesuvianite Tetrahedral Sites.

Site	Z1	Z2	Z3
C.N.	4	4	4
Occupant	Si	Si	Si
Point Sym.	$\bar{4}$	1	1
Wyckoff Not.	4d	16k	16k
Frac.Coord.			
x	3/4	.8192	.9175
y	1/4	.0405	.8496
z	0	.8715	.3644
Distances			
Mean	1.640	1.646	1.628
σ	.000	.026	.020
Poly.Vol.	2.245	2.264	2.211
Q.E.	1.0050	1.0074	1.0010
Ang.Var.	20.6	30.2	4.3
Site Energy	-3889.	-4318.	-4262.
Model Charge	4.0	4.0	4.0

Chapter 6

CHAIN SILICATES

Table 6.1.1. Orthopyroxene Unit Cells.

End-Member	Enstatite	Ferrosilite	Co-Opx
Formula	$Mg_2Si_2O_6$	$Fe_2Si_2O_6$	$Co_2Si_2O_6$
Form.Wt.	200.792	263.862	270.035
Density	3.204	4.002	4.222
Mol Volume	62.676	65.941	63.963
Z	8	8	8
Cryst.Sys.	Orthorhombic	Orthorhombic	Orthorhombic
Cryst.Class	*mmm*	*mmm*	*mmm*
Space Group	*Pbca*	*Pbca*	*Pbca*
Cell Parameters			
a	18.227	18.427	18.296
b	8.819	9.076	8.923
c	5.179	5.237	5.204
Vol.	832.49	875.85	849.58
Ref.	Sasaki et al. (1982b)	Sasaki et al. (1982b)	Sasaki et al. (1982b)

Table 6.1.2. Primitive Clinopyroxene Unit Cells.

End-Member	Enstatite	Ferrosilite	Mn-Cpx
Formula	$Mg_2Si_2O_6$	$Fe_2Si_2O_6$	$Mn_2Si_2O_6$
Form.Wt.	200.792	263.862	262.044
Density	3.188	4.005	3.819
Mol Volume	62.994	65.892	68.608
Z	4	4	4
Cryst.Sys.	Monoclinic	Monoclinic	Monoclinic
Cryst.Class	$2/m$	$2/m$	$2/m$
Space Group	$P\,2_1/c$	$P\,2_1/c$	$P\,2_1/c$
Cell Parameters			
a	9.626	9.7085	9.864
b	8.825	9.0872	9.179
c	5.188	5.2284	5.298
β	108.33	108.43	108.22
Vol.	418.36	437.60	455.64
Ref.	Morimoto et al.(1960)	Burnham (1967)	Tokonami et al. (1979)

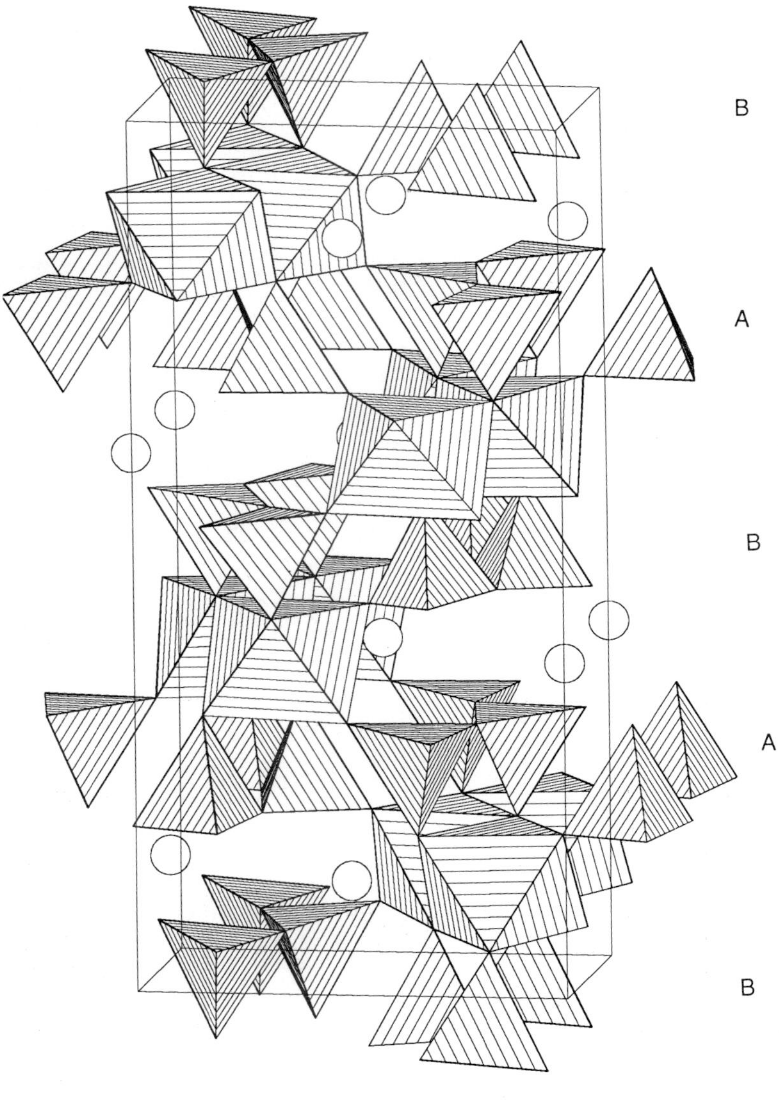

Figure 6.1. Orthoenstatite, $Mg_2Si_2O_6$, perspective c-axis projection, a vertical. Si tetrahedra share a bridging oxygen to form a single infinite chain parallel to c. There are two distinct types of silicate chains, A and B, so that each $b-c$ layer of chains contains only one type of chain. There are two distinct Mg sites, M1 and M2. M1 is a fairly regular octahedron, whereas M2 (circle) is in distorted six-coordination. All cations are in the general position with point symmetry 1.

Table 6.1.3. Orthopyroxene and P-Clinopyroxene M1 sites.

	Orthopyroxene			P-Clinopyroxene		
End-Member	Enstatite	Ferrosilite	Co-Opx	Enstatite	Ferrosilite	Mn-Cpx
C.N.	6	6	6	6	6	6
Occupant	Mg	Fe^{+2}	Co	Mg	Fe^{+2}	Mn
Point Sym.	1	1	1	1	1	1
Wyckoff Not.	8c	8c	8c	4e	4e	4e
Frac.Coord.						
x	.37582	.37574	.37597	.253	.2508	.2510
y	.65379	.65397	.65456	.653	.6533	.6507
z	.86597	.87456	.87202	.220	.2255	.2319
Distances						
O1A	2.151	2.193	2.159	2.022	2.101	2.133
O1A	2.028	2.086	2.047	2.167	2.199	2.255
O1B	2.065	2.128	2.089	2.210	2.200	2.274
O1B	2.172	2.195	2.185	2.007	2.127	2.157
O2A	2.007	2.085	2.060	1.956	2.082	2.114
O2B	2.045	2.122	2.078	2.042	2.114	2.131
Mean	2.078	2.135	2.103	2.067	2.137	2.177
σ	.068	.049	.056	.099	.050	.069
Poly.Vol.	11.83	12.81	12.25	11.61	12.86	13.60
Q.E.	1.0088	1.0088	1.0086	1.0120	1.0083	1.0088
Ang.Var.	26.5	28.7	26.9	31.8	27.1	26.8
Site Energy	-1242.	-1195.	-1221.	-1121.	-1195.	-1165.

Table 6.1.4. Orthopyroxene and P-clinopyroxene M2 sites.

	Orthopyroxene			P-Clinopyroxene		
End-Member	Enstatite	Ferrosilite	Co-Opx	Enstatite	Ferrosilite	Mn-Cpx
C.N.	6	6	6	6	6	6
Occupant	Mg	Fe^{+2}	Co	Mg	Fe^{+2}	Mn
Point Sym.	1	1	1	1	1	1
Wyckoff Not.	8c	8c	8c	4e	4e	4e
Frac.Coord.						
x	.37677	.37775	.37723	.258	.2570	.2535
y	.48698	.48566	.48798	.014	.0142	.0181
z	.35879	.3664	.36164	.193	.2233	.2292
Distances						
O1A	2.088	2.161	2.117	2.121	2.159	2.180
O1B	2.055	2.123	2.072	2.024	2.135	2.162
O2A	2.032	2.023	2.019	2.057	2.032	2.088
O2B	1.994	1.987	1.982	2.048	1.986	2.053
O3A	2.288	2.453	2.388	2.329	2.444	2.472
O3B	2.447	2.589	2.516	2.286	2.587	2.732
Mean	2.151	2.223	2.182	2.144	2.224	2.281
σ	.178	.243	.218	.131	.239	.266
Poly.Vol.	12.46	13.43	12.86	12.53	13.55	14.45
Q.E.	1.0489	1.0700	1.0597	1.0357	1.0641	1.0746
Ang.Var.	140.	181.	159.	109.	164.	192.
Site Energy	-1134.	-1099.	-1119.	-1281.	-1092.	-1063.

Table 6.1.5. Orthopyroxene and P-clinopyroxene T1 (SiA) Sites.

	Orthopyroxene			P-Clinopyroxene		
End-Member	Enstatite	Ferrosilite	Co-Opx	Enstatite	Ferrosilite	Mn-Cpx
C.N.	4	4	4	4	4	4
Occupant	Si	Si	Si	Si	Si	Si
Point Sym.	1	1	1	1	1	1
Wyckoff Not.	8c	8c	8c	4e	4e	4e
Frac.Coord.						
x	.27172	.27231	.27200	.043	.0447	.0431
y	.34162	.3390	.34021	.342	.3386	.3385
z	.05040	.0494	.0520	.294	.2924	.2862
Distances						
O1A	1.611	1.612	1.617	1.655	1.599	1.614
O2A	1.587	1.604	1.597	1.605	1.603	1.603
O3A	1.665	1.652	1.654	1.626	1.629	1.648
O3A	1.646	1.636	1.635	1.674	1.658	1.666
Mean	1.628	1.626	1.625	1.640	1.622	1.633
σ	.035	.022	.024	.030	.027	.029
Poly.Vol.	2.182	2.181	2.180	2.249	2.173	2.216
Q.E.	1.0099	1.0075	1.0076	1.0044	1.0058	1.0057
Ang.Var.	39.8	31.1	30.7	16.7	24.6	23.2
Site Energy	-4406.	-4427.	-4429.	-4375.	-4430.	-4410.

Table 6.1.6. Orthopyroxene and P-clinopyroxene T2 (SiB) Sites.

	Orthopyroxene			P-Clinopyroxene		
End-Member	Enstatite	Ferrosilite	Co-Opx	Enstatite	Ferrosilite	Mn-Cpx
C.N.	4	4	4	4	4	4
Occupant	Si	Si	Si	Si	Si	Si
Point Sym.	1	1	1	1	1	1
Wyckoff Not.	8c	8c	8c	4e	4e	4e
Frac.Coord.						
x	.47358	.47315	.47259	.553	.5538	.5498
y	.33734	.3342	.33561	.839	.8339	.8354
z	.79827	.7893	.79343	.236	.2393	.2448
Distances						
O1B	1.618	1.619	1.622	1.621	1.614	1.616
O2B	1.588	1.603	1.599	1.587	1.612	1.598
O3B	1.678	1.667	1.665	1.729	1.685	1.670
O3B	1.675	1.663	1.666	1.710	1.629	1.666
Mean	1.640	1.638	1.638	1.662	1.635	1.637
σ	.044	.032	.034	.068	.034	.036
Poly.Vol.	2.247	2.241	2.241	2.328	2.228	2.240
Q.E.	1.0054	1.0045	1.0048	1.0044	1.0052	1.0042
Ang.Var.	19.4	17.3	17.6	34.1	19.3	14.7
Site Energy	-4351.	-4354.	-4345.	-4276.	-4367.	-4376.

Table 6.2.1. C-centered Clinopyroxene Unit Cells.

End-Member	Diopside	Hedenbergite	Jadeite	Acmite	Ureyite	Spodumene	Ca-Tschermaks
Formula	$CaMgSi_2O_6$	$CaFeSi_2O_6$	$NaAlSi_2O_6$	$NaFeSi_2O_6$	$NaCrSi_2O_6$	$LiAlSi_2O_6$	$CaAlAlSiO_6$
Form.Wt.	216.560	248.095	202.140	231.005	227.154	186.089	218.125
Density	3.279	3.656	3.341	3.576	3.592	3.176	3.438
Mol Volume	66.039	67.867	60.508	64.606	63.239	58.596	63.445
Z	4	4	4	4	4	4	4
Cryst.Sys.	Monoclinic	Monoclinic	Monoclinic	Monoclinic	Monoclinic	Monoclinic	Monoclinic
Cryst.Class	$2/m$	$2/m$	$2/m$	$2/m$	$2/m$	$2/m$	$2/m$
Space Group	$C2/c$	$C2/c$	$C2/c$	$C2/c$	$C2/c$	$C2/c$	$C2/c$
Cell Parameters							
a	9.746	9.845	9.423	9.658	9.579	9.461	9.609
b	8.899	9.024	8.564	8.795	8.722	8.395	8.652
c	5.251	5.245	5.223	5.294	5.267	5.218	5.274
β	105.63	104.70	107.56	107.42	107.37	110.09	106.06
Vol.	438.58	450.72	401.85	429.06	419.98	389.15	421.35
Ref.	Cameron et al.(1973)	Cameron et al.(1973)	Cameron et al.(1973)	Clark et al.(1969)	Cameron et al.(1973)	Sasaki et al.(1980)	Okamura et al.(1974)

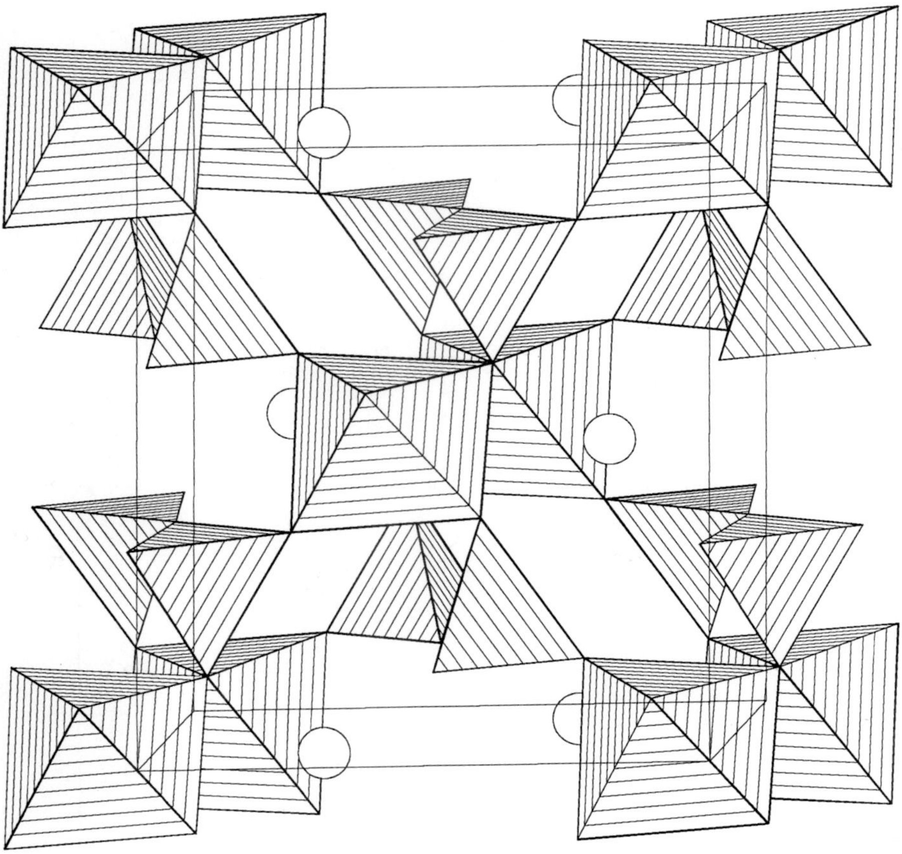

Figure 6.2. Diopside, $CaMgSi_2O_6$, perspective c-axis projection, b horizontal. There is a single Si site in tetrahedral coordination and point symmetry 1. The tetrahedra share bridging oxygens to form a single infinite chain parallel to c with a two-tetrahedron repeat. Mg is in the M1 site, a fairly regular octahedron, and Ca (circle) is in M2, an irregular eight-coordinated site. Both M sites have point symmetry 2.

Table 6.2.3. C-centered Clinopyroxene M1 sites.

End-Member	Diopside	Hedenbergite	Jadeite	Acmite	Ureyite	Spodumene	Ca-Tschermaks
C.N.	6	6	6	6	6	6	6
Occupant	Mg	Fe^{+2}	Al	Fe^{+3}	Cr	Al	Al
Point Sym.	2	2	2	2	2	2	2
Wyckoff Not.	4e	4e	4e	4e	4e	4e	4e
Frac.Coord.							
x	0	0	0	0	0	0	0
y	.9082	.9045	.9058	.8989	.9076	.9066	.90934
z	1/4	1/4	1/4	1/4	1/4	1/4	1/4
Distances							
O1(2)	2.064	2.184	1.995	2.109	2.042	1.997	1.947
O1(2)	2.115	2.141	1.940	2.029	2.010	1.946	2.021
O2(2)	2.050	2.068	1.852	1.936	1.950	1.820	1.872
Mean	2.077	2.131	1.929	2.025	2.001	1.921	1.947
σ	.031	.052	.064	.078	.042	.082	.066
Poly.Vol.	11.85	12.81	9.37	10.87	10.55	9.26	9.64
Q.E.	1.0050	1.0060	1.0152	1.0131	1.0094	1.0150	1.0140
Ang.Var.	17.4	17.4	47.8	41.9	28.3	44.4	44.3
Site Energy	-1284.	-1232.	-2588.	-2424.	-2445.	-2584.	-2587.

Table 6.2.3. C-centered Clinopyroxene M2 Sites.

End-Member	Diopside	Hedenbergite	Jadeite	Acmite	Ureyite	Spodumene	Ca-Tschermaks
C.N.	8	8	8	8	8	6	8
Occupant	Ca	Ca	Na	Na	Na	Li	Ca
Point Sym.	2	2	2	2	2	2	2
Wyckoff Not.	4e	4e	4e	4e	4e	4e	4e
Frac.Coord.							
x	0	0	0	0	0	0	0
y	.3015	.3003	.3005	.2999	.3008	.2752	.31117
z	1/4	1/4	1/4	1/4	1/4	1/4	1/4
Distances							
O1(2)	2.359	2.354	2.356	2.398	2.381	2.111	2.403
O2(2)	2.352	2.342	2.412	2.416	2.389	2.280	2.420
O3(2)	2.561	2.627	2.366	2.430	2.427	2.246	2.469
O3(2)	2.717	2.719	2.741	2.831	2.766		2.549
Mean	2.498	2.511	2.469	2.519	2.491	2.213	2.460
σ	.163	.177	.169	.193	.171	.083	.061
Poly.Vol.	25.76	26.11	24.58	26.30	25.45	10.78	24.52
Q.E.						1.2172	
Ang.Var.						549.	
Site Energy	-957.	-960.	-313.	-308.	-314.	-347.	-975.

Table 6.2.4. C-centered Clinopyroxenes Si Sites.

End-Member	Diopside	Hedenbergite	Jadeite	Acmite	Ureyite	Spodumene	Ca-Tschermaks
C.N.	4	4	4	4	4	4	4
Occupant	Si	Si	Si	Si	Si	Si	Si/Al
Point Sym.	1	1	1	1	1	1	1
Wyckoff Not.	8f	8f	8f	8f	8f	8f	8f
Frac.Coord.							
x	.2862	.2878	.2906	.2905	.2921	.29413	.28802
y	.0933	.0924	.0933	.0894	.0918	.09342	.09693
z	.2293	.2326	.2277	.2351	.2333	.25594	.21337
Distances							
O1	1.602	1.601	1.636	1.628	1.632	1.641	1.693
O2	1.585	1.586	1.594	1.598	1.589	1.586	1.665
O3	1.665	1.666	1.629	1.637	1.639	1.623	1.683
O3	1.687	1.687	1.639	1.646	1.645	1.627	1.701
Mean	1.635	1.635	1.625	1.627	1.626	1.619	1.685
σ	.049	.049	.021	.021	.026	.022	.015
Poly.Vol.	2.221	2.224	2.182	2.201	2.196	2.164	2.425
Q.E.	1.0073	1.0058	1.0062	1.0031	1.0038	1.0050	1.0094
Ang.Var.	28.6	24.9	23.1	13.9	15.9	18.4	35.8
Site Energy	-4395.	-4398.	-4440.	-4442.	-4446.	-4459.	-3435.

Table 6.3.1. Pyroxenoid Unit Cells.

End-Member	Wollastonite	Bustamite	Rhodonite	Pyroxmangite
Formula	$Ca_3Si_3O_9$	$(Ca_{.78}Mn_{.12}Fe_{.10})_3$ Si_3O_9	$Mn_5Si_5O_{15}$	$Mn_7Si_7O_{21}$
Form.Wt.	348.493	358.572	655.111	917.156
Density	2.937	3.116	3.752	3.749
Mol Volume	118.66	115.09	174.62	244.63
Z	4	4	2	2
Cryst.Sys.	Triclinic	Triclinic	Triclinic	Triclinic
Cryst.Class	$\bar{1}$	$\bar{1}$	$\bar{1}$	$\bar{1}$
Space Group	$C\bar{1}$	$I\bar{1}$	$P\bar{1}$	$P\bar{1}$
Cell Parameters				
a	10.104	9.994	7.616	6.721
b	11.054	10.946	11.851	7.603
c	7.305	7.231	6.707	17.455
α	99.53	99.30	92.55	113.18
β	100.56	100.56	94.35	82.27
γ	83.44	83.29	105.67	94.13
Vol.	788.04	764.30	579.84	812.31
Ref.	Ohashi & Finger (1978)	Ohashi & Finger (1978)	Narita et al. (1977)	Narita et al. (1977)

Table 6.3.2a. Pyroxenoid M Sites.

End-Member	Wollastonite			Bustamite			
Site	M1	M2	M3	M1	M2	M3	M4
C.N.	6	6	7	6	6	6	8
Occupant	Ca	Ca	Ca	Ca	Ca	Mn/Fe	Ca
Point Sym.	1	1	1	1	1	$\bar{1}$	$\bar{1}$
Wyckoff Not.	2i	2i	2i	2i	2i	1g	1c
Frac.Coord.							
x	.0212	.0180	.0137	.0222	.0247	0	0
y	.7800	.7803	.4889	.7771	.7766	1/2	1/2
z	.0772	.5717	.2504	.8149	.3133	1/2	0
Distances							
Mean	2.373	2.381	2.414	2.348	2.372	2.195	2.490
σ	.112	.072	.117	.092	.090	.036	.120
Poly.Vol.	16.43	16.61	20.29	16.04	16.24	13.84	24.78
Q.E.	1.0578	1.0558	--	1.0514	1.0637	1.0128	--
Ang.Var.	176.7	177.2	--	161.3	189.6	40.9	--
Site Energy	-1028.	-992.	-1022.	-1036.	-994.	-1141.	-1004.

Table 6.3.2b. Pyroxenoid M Sites (continued).

End-Member	Rhodonite				
Site	M1	M2	M3	M4	M5
C.N.	6	6	6	6	7
Occupant	Mn	Mn	Mn	Mn	Mn
Point Sym.	1	1	1	1	1
Wyckoff Not.	2i	2i	2i	2i	2i
Frac.Coord.					
x	.8290	.6839	.4897	.2962	.0366
y	.8520	.5540	.2693	.9718	.7036
z	.9717	.8712	.8130	.7934	.6519
Distances					
Mean	2.218	2.242	2.223	2.267	2.397
σ	.076	.095	.095	.282	.327
Poly.Vol.	14.13	14.48	13.57	13.84	18.57
Q.E.	1.0208	1.0264	1.0542	1.0939	--
Ang.Var.	65.2	88.8	183.5	246.5	--
Site Energy	-1134.	-1112.	-1080.	-1108.	-1066.

Table 6.3.2c. Pyroxenoid M Sites (continued).

End-Member	Pyroxmangite						
Site	M1	M2	M3	M4	M5	M6	M7
C.N.	6	6	6	6	6	6	6
Occupant	Mn	Mn	Mn	Mn	Mn	Mn	Mn
Point Sym.	1	1	1	1	1	1	1
Wyckoff Not.	2i	2i	2i	2i	2i	2i	2i
Frac.Coord.							
x	.0429	.1711	.0663	.1619	.2649	.7988	.6210
y	.4600	.3318	.4297	.3086	.2268	.8265	.8598
z	.3978	.1873	.8943	.6934	.9903	.5177	.7083
Distances							
Mean	2.233	2.220	2.222	2.222	2.310	2.272	2.284
σ	.079	.092	.091	.091	.277	.294	.225
Poly.Vol.	14.25	14.26	14.34	13.67	12.23	13.88	14.63
Q.E.	1.0196	1.0163	1.0143	1.0479	1.2328	1.0981	1.0645
Ang.Var.	61.3	52.6	44.7	163.8	455.1	257.0	175.9
Site Energy	-1120.	-1126.	-1137.	-1079.	-1051.	-1106.	-1075.

Table 6.3.3a. Pyroxenoid Tetrahedral Sites.

End-Member	Wollastonite			Bustamite		
Site	Si1	Si2	Si3	Si1	Si2	Si3
C.N.	4	4	4	4	4	4
Occupant	Si	Si	Si	Si	Si	Si
Point Sym.	1	1	1	1	1	1
Wyckoff Not.	2i	2i	2i	2i	2i	2i
Frac.Coord.						
x	.2265	.2266	.2260	.2267	.2296	.2209
y	.9585	.9576	.1711	.9640	.9573	.1755
z	.8876	.4540	.2237	.6395	.1983	.9727
Distances						
Mean	1.620	1.620	1.634	1.616	1.617	1.632
σ	.032	.028	.034	.025	.026	.034
Poly.Vol.	2.164	2.164	2.199	2.151	2.154	2.185
Q.E.	1.0061	1.0052	1.0128	1.0050	1.0051	1.0142
Ang.Var.	26.2	22.2	57.9	21.6	21.8	63.9
Site Energy	-4433.	-4409.	-4391.	-4437.	-4428.	-4380.

Table 6.3.3b. Pyroxenoid Tetrahedral Sites (continued).

End-Member	Rhodonite				
Site	Si1	Si2	Si3	Si4	Si5
C.N.	4	4	4	4	4
Occupant	Si	Si	Si	Si	Si
Point Sym.	1	1	1	1	1
Wyckoff Not.	2i	2i	2i	2i	2i
Frac.Coord.					
x	.2212	.2593	.4518	.7443	.9226
y	.1250	.4672	.7340	.0892	.3451
z	.4945	.6366	.7024	.7549	.8483
Distances					
Mean	1.623	1.620	1.617	1.623	1.627
σ	.026	.020	.018	.025	.024
Poly.Vol.	2.155	2.168	2.140	2.181	2.195
Q.E.	1.0115	1.0049	1.0092	1.0037	1.0044
Ang.Var.	51.6	20.5	37.3	14.5	19.1
Site Energy	-4422.	-4422.	-4425.	-4406.	-4365.

Table 6.3.3c. Pyroxenoid Tetrahedral Sites (continued).

End-Member	Pyroxmangite						
Site	Si1	Si2	Si3	Si4	Si5	Si6	Si7
C.N.	4	4	4	4	4	4	4
Occupant	Si	Si	Si	Si	Si	Si	Si
Point Sym.	1	1	1	1	1	1	1
Wyckoff Not.	2i	2i	2i	2i	2i	2i	2i
Frac.Coord.							
x	.2415	.1205	.3189	.7804	.5711	.6774	.4928
y	.8533	.9559	.7563	.1343	.3415	.2374	.4117
z	.5653	.7483	.8375	.9702	.8810	.6902	.5888
Distances							
Mean	1.629	1.630	1.633	1.629	1.626	1.617	1.632
σ	.026	.026	.033	.011	.033	.014	.037
Poly.Vol.	2.209	2.211	2.221	2.203	2.177	2.131	2.195
Q.E.	1.0037	1.0039	1.0040	1.0046	1.0087	1.0116	1.0110
Ang.Var.	13.3	16.5	17.0	19.2	35.5	47.1	48.8
Site Energy	-4389.	-4360.	-4388.	-4422.	-4421.	-4425.	-4383.

Table 6.4.1. Ortho-amphibole Unit Cells.

End-Member	Anthophyllite	Gedrite
Formula	$(Mg_{5.53}Fe_{1.47})$ $Si_8O_{22}(OH)_2$	$(Na_{.5}Ca_{.03})$ $(Fe_{1.1}Mg_{4.5}Al_{1.2})$ $(Al_{1.8}Si_{6.3})$ $O_{22}(OH,F)_2$
Form.Wt.	827.23	827.42
Density	3.111	3.184
Mol Volume	265.88	259.84
Z	4	4
Cryst.Sys.	Orthorhombic	Orthorhombic
Cryst.Class	*mmm*	*mmm*
Space Group	*Pnma*	*Pnma*
Cell Parameters		
a	18.560	18.531
b	18.013	17.741
c	5.2818	5.249
β	90.0	90.0
Vol.	1765.82	1725.65
Reference	Finger (1970)	Papike & Ross (1970)

Table 6.4.2. Ortho-amphibole Octahedral Sites.

End-Member	Anthophyllite				Gedrite			
Site	M1	M2	M3	M4	M1	M2	M3	M4
C.N.	6	6	6	7	6	6	6	6
Occupant	$Mg_{.96}Fe_{.04}$	$Mg_{.97}Fe_{.03}$	$Mg_{.97}Fe_{.03}$	$Mg_{.35}Fe_{.65}$	$Mg_{.88}Fe_{.12}$	$Mg_{.4}Al_{.6}$	$Mg_{.9}Fe_{.1}$	$Mg_{.55}Fe_{.45}$
Point Sym.	1	1	m	1	1	1	m	1
Wyckoff Not.	8d	8d	4c	8d	8d	8d	4c	8d
Frac.Coord.								
x	.12489	.12488	.12579	.12371	.1244	.1248	.1249	.1189
y	.16329	.07317	1/4	-.00982	.1611	.0731	1/4	-.0145
z	.3911	-.1099	-.1089	.3877	.3737	-.1281	-.1248	.3636
Distances								
Mean	2.084	2.076	2.070	2.349	2.093	1.987	2.059	2.185
σ	.032	.049	.011	.374	.042	.042	.035	.138
Poly.Vol.	11.877	11.765	11.595	17.281	11.923	10.379	11.274	12.766
Q.E.	1.0107	1.0094	1.0138	--	1.0171	1.0058	1.0222	1.0622
Ang.Var.	34.7	30.8	44.8	--	54.3	18.6	70.8	202.3
Site Energy	-1053.	-1201.	-1044.	-1071.	-1027.	-1906.	-1054.	-1059.
Model Charge	2.0	2.0	2.0	2.0	2.0	2.6	2.0	2.0

Table 6.4.3. Ortho-amphibole Tetrahedral Sites.

End-Member	Anthophyllite				Gedrite			
Site	T1A	T1B	T2A	T2B	T1A	T1B	T2A	T2B
C.N.	4	4	4	4	4	4	4	4
Occupant	Si	Si	Si	Si	$Si_{.66}Al_{.34}$	$Si_{.62}Al_{.38}$	Si	$Si_{.84}Al_{.16}$
Point Sym.	1	1	1	1	1	1	1	1
Wyckoff Not.	8d	8d	8d	8d	8d	8d	8d	8d
Frac.Coord.								
x	.23039	.01863	.22731	.02469	.2315	.0202	.2278	.0266
y	-.1654	-.16626	-.07956	-.08177	-.1631	-.1645	-.0760	-.0802
z	-.4334	.2760	.0622	-.2227	-.4487	.2971	.0502	-.1985
Distances								
Mean	1.620	1.622	1.624	1.633	1.650	1.656	1.616	1.649
σ	.013	.011	.023	.020	.017	.009	.026	.017
Poly.Vol.	2.183	2.189	2.175	2.221	2.306	2.331	2.140	2.286
Q.E.	1.0002	1.0001	1.0076	1.0046	1.0006	1.0001	1.0087	1.0041
Ang.Var.	0.9	0.3	30.5	18.4	2.3	0.7	33.4	15.5
Site Energy	-4433.	-4403.	-4425.	-4393.	-3769.	-3701.	-4405.	-4062.
Model Charge	4.0	4.0	4.0	4.0	3.66	3.62	4.00	3.84

Table 6.5.1. Clinoamphibole Unit Cells.

End-Member	Cummingtonite	Pargasite	Tremolite	Glaucophane
Formula	$(Fe_{1.7}Mg_{.3})$ $(Mg_{4.4}Fe_{.6})$ Si_8 $O_{22}(OH)_2$	$(Na_{.63}K_{.30})$ $Ca_{2.0}$ $(Fe_{1.10}Mg_{3.25}Al_{.55})$ $(Al_{1.86}Si_{6.14})$ $O_{22}((O,OH)_{1.2}F_{.8})$	$(Na_{.38}K_{.12})$ $(Ca_{1.8}Mg_{.2})$ $Mg_{5.0}$ $(Al_{.23}Si_{7.8})$ $O_{22}(OH_{1.3}F_{.7})$	$(Na_{1.84}Ca_{.20})$ $(Fe_{1.0}Mg_{2.38}Al_{1.58}Ti_{.06})$ $(Al_{.08}Si_{7.92})$ $O_{22}(OH,F)_2$
Form.Wt.	853.72	870.30	824.57	822.03
Density	3.142	3.165	3.010	3.135
Mol Volume	271.68	274.94	273.93	262.25
Z	2	2	2	2
Cryst.Sys.	Monoclinic	Monoclinic	Monoclinic	Monoclinic
Cryst.Class	$2/m$	$2/m$	$2/m$	$2/m$
Space Group	$C2/m$	$C2/m$	$C2/m$	$C2/m$
Cell Parameters				
a	9.51	9.910	9.863	9.541
b	18.19	18.022	18.048	17.740
c	5.33	5.312	5.285	5.295
β	101.92	105.78	104.79	103.667
Volume	902.14	912.96	909.60	870.83
Reference	Fischer (1966)	Robinson et al.(1973b)	Hawthorne & Grundy (1976)	Papike & Clark(1968)

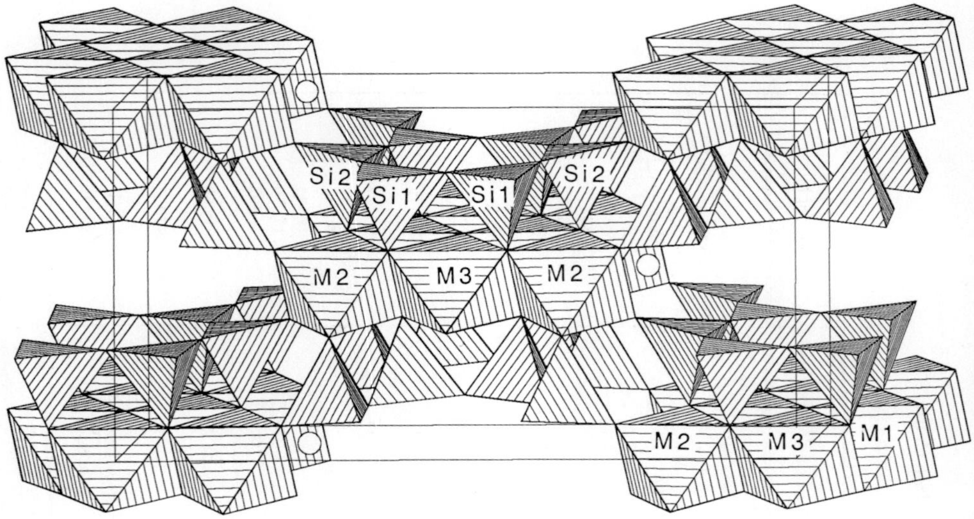

Figure 6.5. Clino-amphibole (glaucophane), $NaFeMg_{2.5}Al_{1.5}Si_8O_{22}(OH)_2$, c-axis perspective projection, a vertical. There are two distinct Si sites, T1 and T2, forming an infinite double silicate chain parallel to c. There are three distinct octahedral sites, M1, M2, and M3. M1 and M2 have point symmetry 2, and of these, M2 is usually smaller and preferentially accepts Al and other smaller cations. M3 has point symmetry $2/m$. M4 (circle) is eight coordinated and similar to the M2 sites of clinopyroxenes. It also has point symmetry 2 and commonly accepts Na or Ca. The A-site (not illustrated) lies in the cavities between the double silicate chains and may accept K, Na, and minor amounts of other large-radius cations. The hydroxyl position is the anion position in the middle of the octahedral band that is not bonded to Si.

Table 6.5.2a. Clinoamphibole Octahedral Sites.

End-Member	Cummingtonite				Pargasite			
Site	M1	M2	M3	M4	M1	M2	M3	M4
C.N.	6	6	6	6	6	6	6	8
Occupant	$Mg_{.84}Fe_{.16}$	$Mg_{.95}Fe_{.05}$	$Mg_{.84}Fe_{.16}$	$Fe_{.84}Mg_{.13}$	Mg,Fe	Mg,Al	Mg,Fe	Ca
Point Sym.	2	2	2/m	2	2	2	2/m	2
Wyckoff Not.	4h	4g	2a	4h	4h	4g	2a	4h
Frac.Coord.								
x	0	0	0	0	0	0	0	0
y	.0872	.1773	0	.2597	.0899	.1766	0	.2802
z	1/2	0	0	1/2	1/2	0	0	1/2
Distances								
Mean	2.100	2.088	2.095	2.297	2.088	2.036	2.078	2.490
σ	.031	.045	.016	.216	.026	.054	.002	.115
Poly.Vol.	12.138	11.969	11.976	11.802	11.855	11.135	11.540	25.87
Q.E.	1.0113	1.0098	1.0161	1.2523	1.0161	1.0079	1.0240	--
Ang.Var.	36.5	32.6	51.8	462.1	50.6	24.3	76.2	--
Site Energy	-1043.	-1192.	-1021.	-1042.	-1288.	-1660.	-1275.	-1008.
Model Charge	2.0	2.0	2.0	2.0	2.0	2.27	2.0	2.0

Table 6.5.2b. Clinoamphibole Octahedral Sites (continued).

End-Member	Tremolite				Glaucophane			
Site	M1	M2	M3	M4	M1	M2	M3	M4
C.N.	6	6	6	8	6	6	6	8
Occupant	Mg	Mg	Mg	Ca,Na	Mg,Fe	Al,Fe	Mg,Fe	Na
Point Sym.	2	2	2/m	2	2	2	2/m	2
Wyckoff Not.	4h	4g	2a	4h	4h	4g	2a	4h
Frac.Coord.								
x	0	0	0	0	0	0	0	0
y	.0883	.1770	0	.2779	.0908	.1807	0	.2772
z	1/2	0	0	1/2	1/2	0	0	1/2
Distances								
Mean	2.074	2.084	2.064	2.518	2.079	1.930	2.094	2.391
σ	.010	.058	.008	.175	.018	.087	.013	--
Poly.Vol.	11.696	11.932	11.498	26.49	11.577	9.435	11.783	--
Q.E.	1.0108	1.0073	1.0133	--	1.0237	1.0121	1.0262	--
Ang.Var.	36.7	25.6	43.8	--	79.4	35.8	84.5	--
Site Energy	-1228.	-1240.	-1216.	-869.	-1229.	-2456.	-1239.	-249.
Model Charge	2.0	2.0	2.0	1.9	2.0	3.0	2.0	1.0

Table 6.5.3. Clinoamphibole Tetrahedral Sites.

End-Member	Cummingtonite		Pargasite		Tremolite		Glaucophane	
Site	T1	T2	T1	T2	T1	T2	T1	T2
C.N.	4	4	4	4	4	4	4	4
Occupant	Si	Si	Si,Al	Si,Al	Si,Al	Si,Al	Si	Si
Point Sym.	1	1	1	1	1	1	1	1
Wyckoff Not.	8j	8j	8j	8j	8j	8j	8j	8j
Frac.Coord.								
x	.2874	.2977	.2799	.2908	.2799	.2882	.2831	.2920
y	.0842	.1688	.0857	.1734	.08424	.17133	.0871	.1730
z	.2746	.7817	.3034	.8141	.2974	.8056	.2931	.8087
Distances								
Mean	1.622	1.633	1.675	1.636	1.627	1.636	1.616	1.640
σ	.009	.016	.015	.023	.014	.042	.004	.034
Poly.Vol.	2.189	2.220	2.407	2.230	2.206	2.229	2.166	2.247
Q.E.	1.0001	1.0039	1.0016	1.0047	1.0014	1.0054	1.0002	1.0047
Ang.Var.	.8	15.9	6.8	19.3	6.2	21.7	0.8	17.6
Site Energy	-4425.	-4404.	-3538.	-4294.	-4269.	-4324.	-4386.	-4208.
Model Charge	4.0	4.0	3.62	3.91	3.97	3.97	4.0	4.0

Table 6.5.4. Clino and Orthoamphibole A-sites.

End-Member	Pargasite	Tremolite		Gedrite
Site	A	A1	A2	A
C.N.	6(12)	10	10	8
Occupant	$Na_{.63}K_{.30}$	Na,K	Na,K	$Na_{.34}$
Point Sym.	1	m	2	m
Wyckoff Not.	8j	4i	4g	4c
Frac.Coord.				
x	.4733	.0450	0	.1151
y	0	1/2	.4897	-1/4
z	.9444	.1030	0	.8533
Distances				
Mean	3.047	2.912	2.946	2.691
σ	.323	.257	.284	.291
Poly.Vol.	66.023	46.304	46.914	25.684
Site Energy	-158.	-2.1	+1.0	-81.
Model Charge	1.0	.08	.08	.34

Table 6.6.1. Aenigmatite Unit Cell.

End-Member	Aenigmatite
Formula	$Na_2Fe^{2+}_{4.90}Ti_{1.10}(Si_{.96}Fe^{3+}_{.04})_6O_{20}$
Form.Wt.	867.49
Density	3.869
Mol Volume	224.21
Z	2
Cryst.Sys.	Triclinic
Cryst.Class	$\bar{1}$
Space Group	$P\bar{1}$
Cell Parameters	
a	10.406
b	10.813
c	8.926
α	104.93
β	96.87
γ	125.32
Vol.	744.52
Ref.	Cannillo et al. (1971)

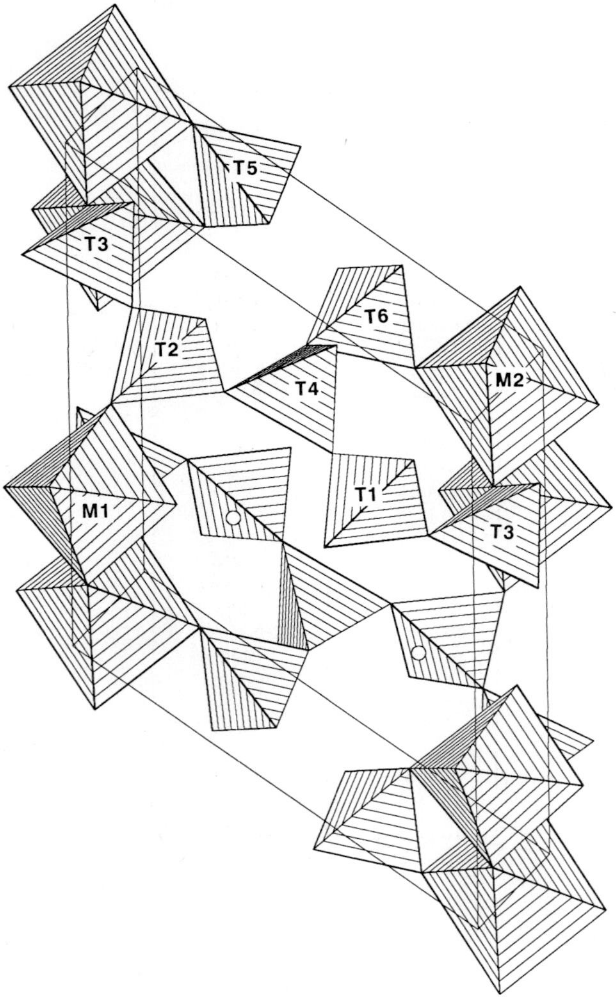

Figure 6.6. Aenigmatite, $Na_2Fe_5TiSi_6O_{20}$, perspective *c*-axis projection, *b* vertical. There are two *Na* sites (circles) in seven-coordination. There are seven octahedral sites containing *Fe* and *Ti*, but most of the *Ti* is in M7. Some of the octahedra are omitted from the drawing to show the six independent silicate tetrahedra forming branched chains.

Table 6.6.2. Aenigmatite Alkali Sites.

End-Member	Aenigmatite	
Site	Na1	Na2
C.N.	7	7
Point Sym.	1	1
Wyckoff Not.	2i	2i
Frac.Coord.		
x	.2089	.6607
y	.6298	.6117
z	.3893	.3741
Distances		
Mean	2.487	2.515
σ	.092	.129
Poly.Vol.	22.520	23.407
Site Energy	-348.	-308.
Model Charge	1.0	1.0

Table 6.6.3. Aenigmatite Octahedral Sites.

End-Member	Aenigmatite						
Site	M1	M2	M3	M4	M5	M6	M7
C.N.	6	6	6	6	6	6	6
Occupant	$Fe_{.89}Ti_{.11}$	$Fe_{.87}Ti_{.13}$	$Fe_{.95}Ti_{.05}$	$Fe_{.76}Ti_{.24}$	$Fe_{.90}Ti_{.10}$	Fe	$Fe_{.41}Ti_{.59}$
Point Sym.	1	1	1	1	1	1	1
Wyckoff Not.	1b	1c	2i	2i	2i	2i	2i
Frac.Coord.							
x	0	0	.3214	.7655	.0961	.5959	.9970
y	0	1/2	.8528	.8199	.9392	.9432	.7434
z	1/2	0	.1779	.1511	.0530	.0661	.2577
Distances							
Mean	2.100	2.107	2.129	2.139	2.157	2.168	1.976
σ	.060	.070	.108	.077	.027	.061	.106
Poly.Vol.	12.112	12.200	12.452	12.657	12.845	12.420	10.189
Q.E.	1.0141	1.0160	1.0241	1.0219	1.0275	1.0091	1.0087
Ang.Var.	43.3	48.7	70.1	68.3	89.8	28.2	22.1
Site Energy	-1366.	-1503.	-1280.	-1538.	-1252.	-1067.	-2815.
Model Charge	2.00	2.26	2.10	2.48	2.20	2.00	3.18

Table 6.6.4. Aenigmatite Tetrahedral Sites.

End-Member	Aenigmatite					
Site	T1	T2	T3	T4	T5	T6
C.N.	4	4	4	4	4	4
Occupant	Si	Si	$Si_{.90}Fe_{.10}$	$Si_{.95}Fe_{.05}$	$Si_{.94}Fe_{.06}$	$Si_{.95}Fe_{.05}$
Point Sym.	1	1	1	1	1	1
Wyckoff Not.	2i	2i	2i	2i	2i	2i
Frac.Coord.						
x	.4768	.9864	.7921	.2772	.6487	.3528
y	.2345	.2363	.3435	.3382	.9448	.5588
z	.3313	.3466	.2416	.2252	.4447	.0501
Distances						
Mean	1.646	1.629	1.647	1.625	1.649	1.630
σ	.031	.021	.027	.026	.013	.034
Poly.Vol.	2.274	2.198	2.267	2.188	2.292	2.218
Q.E.	1.0047	1.0067	1.0076	1.0041	1.0032	1.0017
Ang.Var.	19.6	28.8	30.9	17.7	13.4	6.7
Site Energy	-4272.	-4396.	-4249.	-4178.	-4306.	-4340.
Model Charge	4.0	4.0	3.90	3.95	3.94	3.95

Chapter 7

LAYER SILICATE MINERALS

Table 7.1.1. Talc and Pyrophyllite Unit Cells.

End-Member	Talc	Pyrophyllite
Formula	$Mg_3Si_4O_{10}(OH)_2$	$Al_2Si_4O_{10}(OH)_2$
Form.Wt.	379.289	360.316
Density	2.776	2.814
Mol Volume	136.654	128.036
Z	2	2
Cryst.Sys.	Triclinic	Triclinic
Cryst.Class	$\bar{1}$	$\bar{1}$
Space Group	$C\bar{1}$	$C\bar{1}$
Cell Parameters		
a	5.290	5.160
b	9.173	8.966
c	9.460	9.347
α	90.46	91.18
β	98.68	100.46
γ	90.09	89.64
Vol.	453.77	425.16
Ref.	Perdikatsis & Burzlaff (1981)	Lee & Guggenheim (1981)

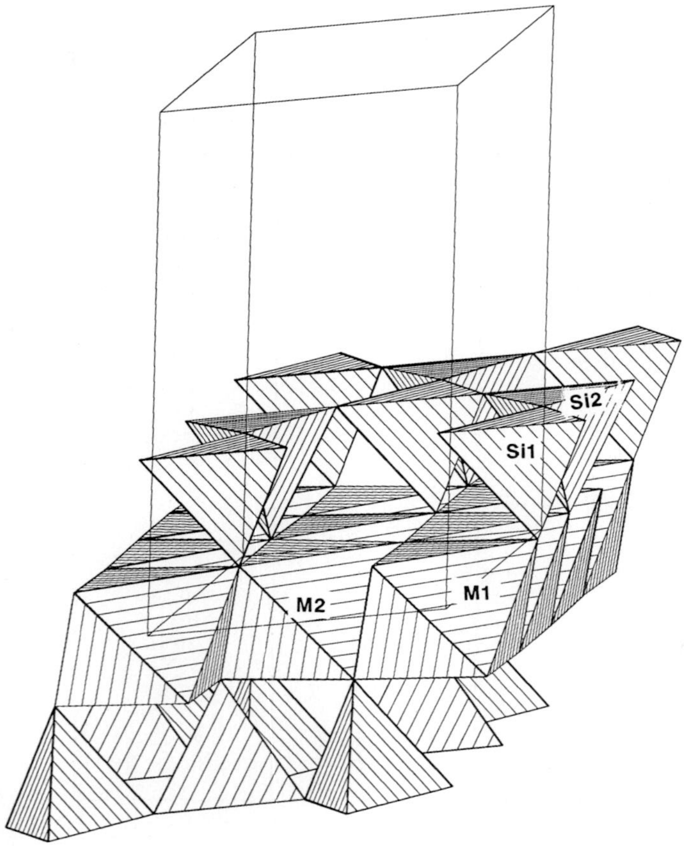

Figure 7.1a. Talc, $Mg_3Si_4O_{10}(OH)_2$, perspective *b*-axis projection, *a* horizontal. There is a complete brucite-like trioctahedral sheet of *Mg* octahedra. There are two distinct octahedral sites; M1 has point symmetry $\bar{1}$ and M2 point symmetry 1. There are two distinct silicate tetrahedra, both in the general position. The hydroxyl is in the octahedral sheet at the anion position not bonded to *Si*. There is no interlayer cation.

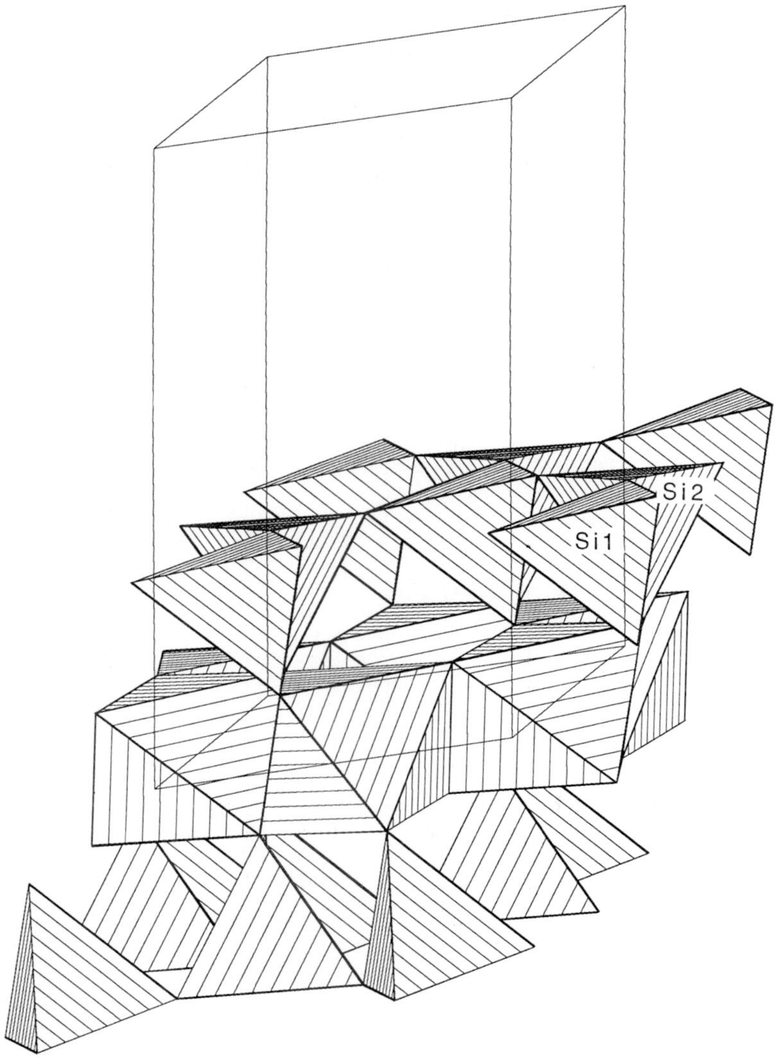

Figure 7.1b. Pyrophyllite, $Al_2Si_4O_{10}(OH)_2$ perspective b-axis projection, a horizontal. All Al sites are equivalent and octahedrally coordinated with point symmetry 1 to form a gibbsite-like dioctahedral sheet. Infinite sheets of silicate tetrahedra attach to either side of the dioctahedral sheet to form an infinite T-O-T sandwich. The hydroxyl is in the octahedral sheet at the anion position not bonded to Si. There is no interlayer cation.

Table 7.1.2. Talc and Pyrophyllite Octahedral Sites.

End-Member	Talc		Pyrophyllite
Site	M1	M2	Al
C.N.	6	6	6
Occupant	Mg	Mg	Al
Point Sym.	$\bar{1}$	1	1
Wyckoff Not.	1a	2i	2i
Frac.Coord.			
x	0	.5001	.4995
y	0	.8333	.16705
z	0	.9999	-.00008
Distances			
Cation-Ox	2.052(2)	2.057	1.926
	2.081(2)	2.054	1.922
	2.080(2)	2.076	1.921
		2.080	1.926
		2.079	1.889
		2.078	1.888
Mean	2.071	2.071	1.912
σ	.015	.012	.018
Poly.Vol.	11.697	11.687	9.069
Q.E.	1.0087	1.0086	1.0183
Ang.Var.	28.6	28.6	66.4
Site Energy	-1138.	-1138.	-2436.

Table 7.1.3. Talc and Pyrophyllite Tetrahedral Sites.

End-Member	Talc		Pyrophyllite	
Site	Si1	Si2	Si1	Si2
C.N.	4	4	4	4
Occupant	Si	Si	Si	Si
Point Sym.	1	1	1	1
Wyckoff Not.	2i	2i	2i	2i
Frac.Coord.				
x	.2453	.2459	.7449	.7595
y	.5026	.8359	-.00303	.32577
z	.2909	.2911	.29169	.29230
Distances				
Cation-Ox	1.621	1.621	1.632	1.634
	1.623	1.622	1.616	1.614
	1.623	1.624	1.618	1.616
	1.625	1.623	1.602	1.607
Mean	1.623	1.622	1.617	1.618
σ	.002	.001	.012	.012
Poly.Vol.	2.194	2.191	2.168	2.171
Q.E.	1.0000	1.0000	1.0005	1.0004
Ang.Var.	.2	.3	2.0	1.7
Site Energy	-4376.	-4379.	-4393.	-4401.

Table 7.2.1. Trioctahedral Mica Unit Cells.

End-Member Polytype	Annite 1M	F-Phlogopite 1M	Phlogopite 1M	Phlogopite-N 1M	Lepidolite $2M_2$	Lepidolite $2M_1$	Lepidolite $1M$	Zinnwaldite $1M$
Formula	$(K_{.88}Na_{.07}Ca_{.03})$ $Fe_{2.4}Ti_{.22}Mg_{.12}$ $Mn_{.05}Al_{.09}$ $Al_{1.2}Si_{2.8}O_{10}$ $(OH)_{1.4}F_{.22}Cl_{.05}$	K Mg_3 $AlSi_3$ $O_{9.9}$ $(OH)_{.16}F_{1.84}$	$(K_{.77}Na_{.16}Ba_{.05})$ $Mg_{3.0}$ $(Al_{1.05}Si_{2.95})$ O_{10} $(OH)_{.7}F_{1.3}$	$(K_{.9}Na_{.02})$ $Mg_{2.7}Fe_{.16}Al_{.08}$ $(Al_{1.09}Si_{2.91})$ O_{10} $(OH)_{.97}F_{1.13}$	$(K_{.9}Na_{.1}Rb_{.1})_2$ $(Al_{.6}Li_{.4})_4$ $(Li_{.95}Al_{.05})_2$ $(Al_{1.3}Si_{6.7})$ $O_{20}(OH)_{.94}F_{3.06}$	$(K_{.8}Na_{.05}Rb_{.06})_2$ $(Al_{.6}Li_{.4})_4$ $(Li_{.9})_2$ $(Al_{.4}Si_{3.6})_2$ $O_{20}((OH)_{.24}F_{.75})_4$	$(K_{.9}Na_{.06}Rb_{.05})_2$ $(Al_{1.3}Li_{1.7})_2$ $(Al_{.64}Si_{3.4})_2$ $O_{20}((OH)_{.25}F_{.75})_4$	$(K_{.9}Na_{.05})_2$ $(Al_{1.05}Fe_{.93}Li_{.67})_2$ $(Al_{.91}Si_{3.09})_2$ $O_{20}((OH)_{.4}F_{.6})_4$
Form.Wt.	490.67	421.268	421.831	422.652	790.04	771.01	793.87	857.99
Density	3.215	2.875	2.872	2.876	2.791	2.724	2.825	2.986
Mol Volume	152.63	146.538	146.869	146.967	283.19	283.03	281.00	289.10
Z	2	2	2	2	2	2	1	1
Cryst.Sys.	Monoclinic	Monoclinic	Monoclinic	Monoclinic	Monoclinic	Monoclinic	Monoclinic	Monoclinic
Cryst.Class	$2/m$	$2/m$	$2/m$	$2/m$	$2/m$	$2/m$	$2/m$	2
Space Group	$C2/m$	$C2/m$	$C2/m$	$C2/m$	$C2/c$	$C2/c$	$C2/m$	$C2$
Cell Parameters								
a	5.3860	5.308	5.3078	5.3141	9.04	5.209	5.20	5.296
b	9.3241	9.183	9.1901	9.2024	5.22	9.053	9.01	9.140
c	10.2683	10.139	10.1547	10.1645	20.210	20.185	10.09	10.096
β	100.63	100.07	100.08	100.05	99.583	99.125	99.28	100.83
Vol.	506.82	486.60	487.69	488.02	940.38	939.82	466.6	480.00
Ref.	Hazen & Burnham(1973)	McCauley et al.(1973)	Hazen & Burnham(1973)	Joswig (1972)	Sartori et al.(1973)	Sartori (1977)	Sartori (1976)	Guggenheim & Bailey(1977)

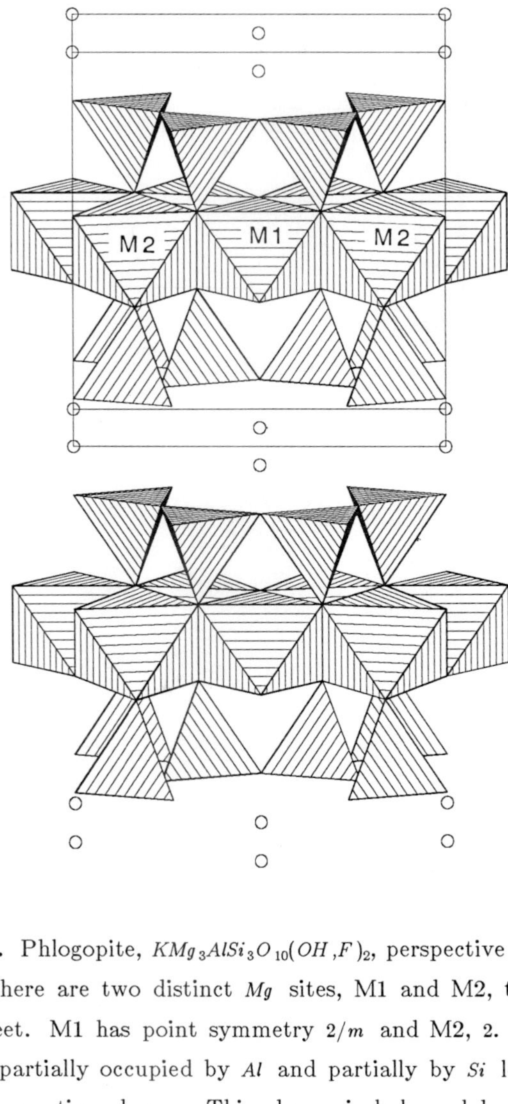

Figure 7.2. Phlogopite, $KMg_3AlSi_3O_{10}(OH,F)_2$, perspective a-axis projection, b horizontal. There are two distinct Mg sites, M1 and M2, that make up the trioctahedral sheet. M1 has point symmetry $2/m$ and M2, 2. There is a single tetrahedral site partially occupied by Al and partially by Si leaving the T-O-T unit with a net negative charge. This charge is balanced by K (circle) in the interlayer positions.

Table 7.2.2. Trioctahedral Mica M Sites.

End-Member Polytype	Annite 1*M*		F-Phlogopite 1*M*		Phlogopite 1*M*		Phlogopite-N 1*M*		Lepidolite $2M_2$		Lepidolite $2M_1$		Lepidolite 1*M*		Zinnwaldite 1*M*		
Site	M1	M2	M1	M2	M1	M2	M1	M2	M1	M2	M1	M2	M1	M2	M1	M2	M3
C.N.	6	6	6	6	6	6	6	6	6	6	6	6	6	6	6	6	6
Occupant	Fe	Fe	Mg	Mg	Mg	Mg	Mg	Mg	Al,Li	Li	Al,Li	Li	$Li_{.4}Al_{.6}$	Li	Fe,Li	Al	Fe,Li
Point Sym.	$2/m$	2	$2/m$	2	$2/m$	2	$2/m$	2	$\bar{1}$	1	$\bar{1}$	1	2	$2/m$	2	2	2
Wyckoff Not.	2d	4h	2d	4h	2d	4h	2d	4h	8f	4d	8f	4d	4h	2c	2b	2b	2b
Frac.Coord.																	
x	0	0	0	0	0	0	0	0	.0857	1/4	.2550	1/4	0	0	0	0	1/2
y	1/2	.8332	1/2	.8306	1/2	0.8315	1/2	0.8312	.2583	3/4	.0851	3/4	0.3283	0	-.0069	.3217	.1631
z	1/2	1/2	1/2	1/2	1/2	1/2	1/2	1/2	.0003	0	.0001	0	1/2	1/2	1/2	1/2	1/2
Distances																	
Mean	2.121	2.101	2.061	2.064	2.063	2.065	2.065	2.061	1.980	2.124	1.977	2.122	1.972	2.113	2.136	1.885	2.135
σ	.001	.014	.025	.022	.025	.026	.034	.012	.017	.026	.051	.049	.018	.005	.006	.018	.019
Poly.Vol.	12.533	12.210	11.475	11.515	11.505	11.523	11.501	11.447	10.180	12.329	10.108	12.243	12.147	10.045	12.425	8.896	12.407
Q.E.	1.0095	1.0081	1.0129	1.0127	1.012	1.012	1.014	1.013	1.0115	1.0237	1.0130	1.0270	1.0237	1.0121	1.0299	1.0025	1.0302
Ang.Var.	31.7	26.9	38.7	40.9	38.5	39.7	44.0	42.5	39.4	74.1	41.9	82.8	74.6	41.2	99.5	8.5	99.9
Site Energy	-1170.	-1195.	-1327.	-1323.	-1289.	-1285.	-1287.	-1288.	-1436.	-348.	-1283.	-176.	-1443.	-329.	-585.	-2340.	-587.
Model Charge	2.0	2.0	2.0	2.0	2.0	2.0	2.0	2.0	2.26	1.00	2.24	1.00	2.26	1.00	1.48	3.00	1.48

Table 7.2.3. Trioctahedral Mica Tetrahedral Sites.

End-Member Polytype	Annite 1*M*	F-Phlogopite 1*M*	Phlogopite 1*M*	Phlogopite-N 1*M*	Lepidolite 2M_2		Lepidolite 2M_1		Lepidolite 1*M*	Zinnwaldite 1*M*	
Site	T	T	T	T	T1	T2	T1	T2	T	T1	T11
C.N.	4	4	4	4	4	4	4	4	4	4	4
Occupant	$Al_{.30}Si_{.70}$	$Al_{.25}Si_{.75}$	$Al_{.26}Si_{.74}$	$Al_{.27}Si_{.73}$	$Al_{.16}Si_{.84}$	$Al_{.16}Si_{.84}$	$Al_{.15}Si_{.85}$	$Al_{.06}Si_{.94}$	$Al_{.16}Si_{.84}$	$Al_{.23}Si_{.77}$	$Al_{.19}Si_{.81}$
Point Sym.	1	1	1	1	1	1	1	1	1	1	1
Wyckoff Not.	8j	8j	8j	8j	8f	8f	8f	8f	8f	8f	8f
Frac.Coord.											
x	.5703	.5751	.5752	.5755	.2937	.1251	.4614	.4558	.0810	.0745	.5844
y	.1665	.1663	.1668	.16665	.0935	.5865	.9244	.2554	.1685	.1688	.3323
z	.2246	.2245	.2254	.2257	.1338	.1338	.1340	.1341	.2320	.2276	.2275
Cation-Ox											
Mean	1.660	1.642	1.649	1.652	1.630	1.627	1.643	1.628	1.631	1.649	1.642
σ	.006	.006	.002	.006	.018	.012	.023	.027	.018	.004	.003
Poly.Vol.	2.345	2.270	2.301	2.312	2.217	2.206	2.266	2.204	2.217	2.299	2.270
Q.E.	1.0003	1.0002	1.0003	1.0002	1.0017	1.0019	1.0037	1.0038	1.0025	1.0009	1.0009
Ang.Var.	1.2	0.7	1.5	1.8	7.9	8.5	14.2	14.3	10.5	3.6	3.7
Site Energy	-3830.	-4193.	-4297.	-4289.	-4019.	-4026.	-4163.	-4369.	-3926.	-3933.	-4022.
Model Charge	3.70	3.75	3.74	3.73	3.84	3.84	3.85	3.94	3.84	3.77	3.81

Table 7.2.4. Trioctahedral Mica Alkali Sites.

End-Member Polytype	Annite 1*M*	F-Phlogopite 1*M*	Phlogopite 1*M*	Phlogopite-N 1*M*	Lepidolite $2M_2$	Lepidolite $2M_1$	Lepidolite 1*M*	Zinnwaldite 1*M*
C.N.	12	12	12	12	12	12	12	12
Occupant	K,Na	K	K,Na	K	K	K	K	K
Point Sym.	$2/m$	$2/m$	$2/m$	$2/m$	2	2	$2/m$	2
Wyckoff Not.	2a	2a	2a	2a	4e	4e	2b	2a
Frac.Coord.								
x	0	0	0	0	0	0	0	0
y	0	0	0	0	.0916	.0906	1/2	.5028
z	0	0	0	0	1/4	1/4	0	0
Distances								
Mean	3.179	3.140	3.140	3.142	3.119	3.111	3.105	3.126
σ	.048	.139	.179	.183	.153	.169	.172	.143
Poly.Vol.	64.443	77.367	77.967	78.068	60.85	67.89	62.90	63.83
Site Energy	-220.	-171.	-240.	-240.	-251.	-371.	-321.	-291.
Model Charge	1.0	1.0	1.0	1.0	1.0	1.0	1.0	1.0

Table 7.3.3a. Dioctahedral Mica Tetrahedral Sites.

End-Member Polytype	Muscovite $2M_1$		Paragonite $2M_1$		Margarite $2M_1$			
Site	T1	T2	T1	T2	T1	T2	T11	T22
C.N.	4	4	4	4	4	4	4	4
Occupant	$Al_{.25}Si_{.75}$	$Al_{.25}Si_{.75}$	$Al_{.27}Si_{.73}$	$Al_{.27}Si_{.73}$	$Al_{.85}Si_{.15}$	$Si_{.85}Al_{.15}$	$Si_{.91}Al_{.09}$	$Al_{.79}Si_{.21}$
Point Sym.	1	1	1	1	1	1	1	1
Wyckoff Not.	8f	8f	8f	8f	8f	8f	8f	8f
Frac.Coord.								
x	.4646	.4516	.9528	.4401	.4623	.4544	.5336	.5467
y	.9291	.2581	.4288	.2578	.9264	.2567	.0739	.7440
z	.13553	.13559	.1409	.1409	.1422	.1445	.8551	.8576
Distances								
Mean	1.645	1.644	1.653	1.652	1.747	1.633	1.622	1.736
σ	.007	.003	.004	.007	.021	.006	.007	.016
Poly.Vol.	2.279	2.278	2.314	2.312	2.729	2.231	2.186	2.681
Q.E.	1.0012	1.0010	1.0007	1.0007	1.0017	1.0009	1.0006	1.0014
Ang.Var.	5.0	4.1	2.9	3.0	7.3	3.6	2.4	5.5
Site Energy	-3873.	-3873.	-3797.	-3798.	-2672.	-4056.	-4170.	-2758.
Model Charge	3.75	3.75	3.73	3.73	3.15	3.85	3.91	3.21

Table 7.3.1. Dioctahedral Mica Unit Cells.

End-Member Polytype	Muscovite $2M_1$	Paragonite $2M_1$	Margarite $2M_2$
Formula	K_2 Al_4 Si_6Al_2 $O_{20}(OH)_4$	$(Na_{.92}K_{.04}Ca_{.02})_2$ $(Al_{1.99}Fe_{.03}Mg_{.01})_2$ $Si_{5.88}Al_{2.12}$ $O_{20}(OH)_4$	$(K_{.01}Na_{.19}Ca_{.81})_2$ Al_4 $(Al_{1.89}Si_{2.11})_2$ $O_{20}(OH)_4$
Form.Wt.	796.627	768.62	791.82
Density	2.834	2.909	3.061
Mol Volume	281.14	264.26	258.66
Z	2	2	2
Cryst.Sys.	Monoclinic	Monoclinic	Monoclinic
Cryst.Class	$2/m$	$2/m$	m
Space Group	$C2/c$	$C2/c$	Cc
Cell Parameters			
a	5.1918	5.128	5.1038
b	9.0153	8.898	8.8287
c	20.0457	19.287	19.148
β	95.735	94.35	95.46
Vol.	933.56	877.51	858.89
Ref.	Rothbauer (1971)	Lin & Bailey(1984)	Guggenheim & Bailey(1975,1978)

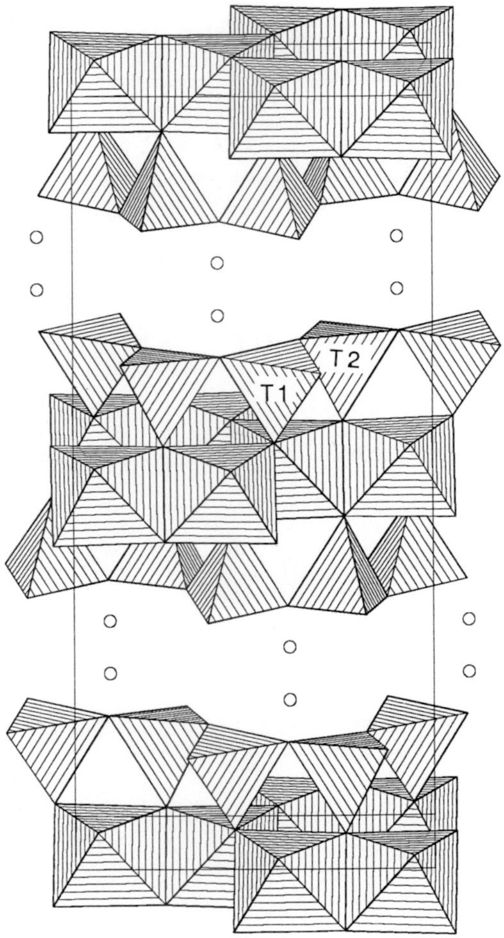

Figure 7.3. Muscovite ($2M_1$), $KAl_2(AlSi_3)O_{10}(OH)_2$, perspective *a*-axis projection, *b* horizontal. There is a single *Al* octahedron with point symmetry 1 forming the dioctahedral sheet. There are two distinct tetrahedra, T1 and T2, over which *Al* and *Si* appear to be disordered in a one-to-three ratio, leaving a net negative charge on the T-O-T unit. This charge is balanced by *K* (circles) in the interlayer.

Table 7.3.2. Dioctahedral Mica Octahedral Sites.

End-Member Polytype	Muscovite $2M_1$	Paragonite $2M_1$	Margarite $2M_1$	
Site	M1	M2	M2	M3
C.N.	6	6	6	6
Occupant	Al	Al	Al	Al
Point Sym.	1	1	1	1
Wyckoff Not.	8f	8f	4a	4a
Frac.Coord.				
x	.2502	.2499	.7449	.2488
y	.0835	.0832	.9177	.0858
z	.00008	-.00002	.9996	-.0003
Distances				
Mean	1.930	1.908	1.903	1.915
σ	.012	.016	.034	.054
Poly.Vol.	9.355	9.040	9.006	9.151
Q.E.	1.0162	1.0162	1.0136	1.0160
Ang.Var.	59.0	59.0	47.9	55.0
Site Energy	-2403.	-2495.	-2702.	-2674.
Model Charge	3.0	3.0	3.0	3.0

Table 7.3.4. Dioctahedral Mica Alkali Sites.

End-Member Polytype	Muscovite $2M_1$	Paragonite $2M_1$	Margarite $2M_1$
C.N.	10	6	6
Occupant	K	Na	Ca
Point Sym.	2	2	1
Wyckoff Not.	4e	4e	4a
Frac.Coord.			
x	0	0	0
y	.0980	.0941	.0933
z	1/4	1/4	1/4
Distances			
Mean	3.031	2.624	2.455
σ	.224	.102	.032
Poly.Vol.	51.27	23.95	19.66
Q.E.		1.0051	1.0021
Ang.Var.		9.4	6.6
Site Energy	-298.	-314.	-920.
Model Charge	1.0	1.0	2.0

Table 7.4.1. Clay Mineral Unit Cells.

End-Member	Nacrite	Dickite	Kaolinite	Amesite	Lizardite
Formula	$Al_2Si_2O_5(OH)_4$	$Al_2Si_2O_5(OH)_4$	$Al_2Si_2O_5(OH)_4$	$Mg_{1.7}Al_{.95}Fe^{2+}_{.315}$ $Si_{1.09}Al_{.92}O_5(OH)_4$	$Mg_{2.79}Fe^{2+}_{.04}Fe^{3+}_{.10}Al_{.07}$ $Si_{1.83}Al_{.17}O_5(OH)_4$
Form.Wt.	258.162	258.162	258.162	288.02	281.55
Density	2.602	2.619	2.601	2.778	2.625
Mol Volume	99.221	98.578	99.236	103.69	107.26
Z	4	4	2	4	1
Cryst.Sys.	Monoclinic	Monoclinic	Triclinic	Triclinic	Trigonal
Cryst.Class	m	m	1	1	31m
Space Group	Cc	Cc	P1	C1	P31m
Cell Parameters					
a	8.909	5.1375	5.153	5.319	5.332
b	5.146	8.9178	8.941	9.208	5.332
c	15.697	14.389	7.403	14.060	7.233
α	90.0	90.0	91.692	90.01	90.0
β	113.70	96.74	104.860	90.27	90.0
γ	90.0	90.0	89.822	89.96	120.0
Vol.	658.95	654.68	329.52	688.61	178.09
Ref.	Blount et al. (1969)	Joswig & Drits (1986)	Suitch & Young (1983)	Hall & Bailey (1979)	Mellini (1982)

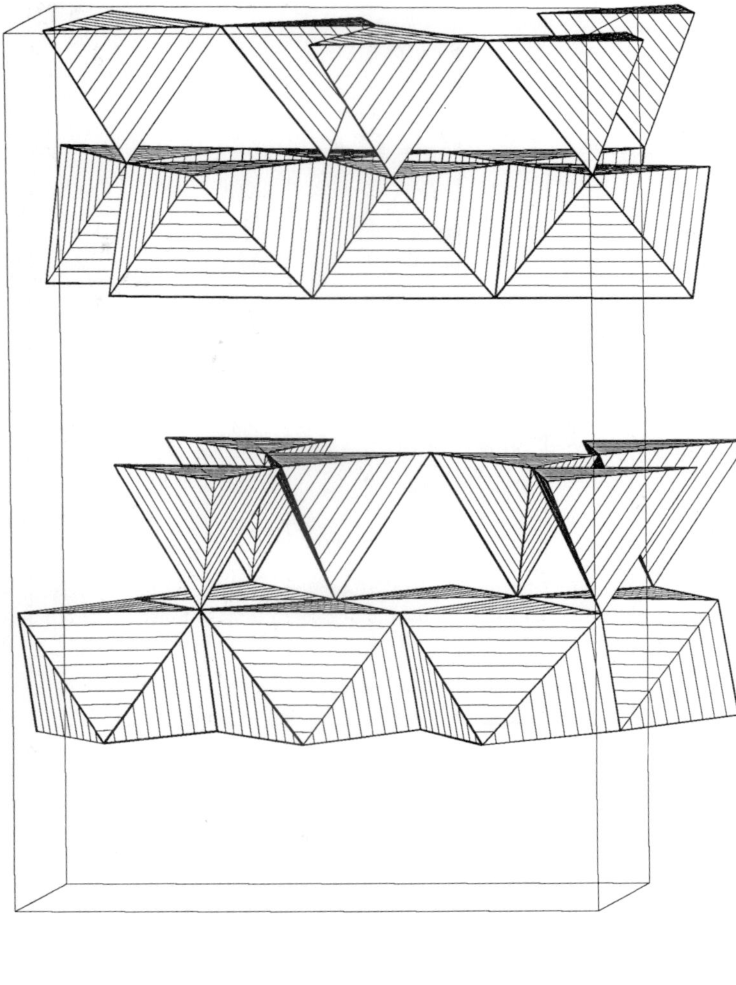

Figure 7.4. Amesite, $(Mg_{1.7}Al_{1.0}Fe_{.3})(Si_{1.1}Al_{.9})O_5(OH)_4$ perspective *a*-axis projection, *b* horizontal. There are two distinct infinite trioctahedral sheets each composed of three distinct octahedra, for a total of six different octahedral sites. Amesite is ordered with trivalent cations in one of the three octahedra in each sheet. There are also two distinct silicate sheets each composed of an *Al* and a *Si* tetrahedron. There are no cations in the interlayer.

Table 7.4.2a. Clay Mineral Octahedral Sites.

End-Member	Nacrite		Dickite		Kaolinite			
Site	Al(1)	Al(2)	Al(1)	Al(2)	Al(1)	Al(2)	Al(3)	Al(4)
C.N.	6	6	6	6	6	6	6	6
Occupant	Al	Al	Al	Al	Al	Al	Al	Al
Point Sym.	1	1	1	1	1	1	1	1
Wyckoff Not.	4a	4a	4a	4a	4a	4a	4a	4a
Frac.Coord.								
x	.1582	.5004	-.0854	.4193	.367	.339	.846	.847
y	.3389	.3248	.25257	.41760	.490	.824	-.006	.314
z	.2202	.2199	.23162	.23092	.443	.425	.442	.457
Distances								
Mean	1.915	1.920	1.900	1.900	1.916	1.885	1.948	1.910
σ	.075	.034	.063	.056	.114	.079	.095	.134
Poly.Vol.	9.109	9.192	8.869	8.877	9.022	8.676	9.534	8.920
Q.E.	1.0200	1.0178	1.0211	1.0206	1.0295	1.0205	1.0241	1.0313
Ang.Var.	66.2	63.2	73.2	72.2	91.9	64.4	76.8	93.0
Site Energy	-2541.	-2536.	-2560.	-2523.	-2663.	-2648.	-2493.	-2640.

Table 7.4.2b. Clay Mineral Octahedral Sites (continued).

End-Member	Amesite						Lizardite
Site	M1	M2	M3	M11	M22	M33	M1
C.N.	6	6	6	6	6	6	6
Occupant	Mg	Mg	Al	Al	Mg	Mg	Mg
Point Sym.	1	1	1	1	1	1	m
Wyckoff Not.	1a	1a	1a	1a	1a	1a	3c
Frac.Coord.							
x	.1668	.6678	.6721	.3337	.3294	.8305	.3327
y	.1710	.0013	.3372	.3359	.0003	.1650	0
z	.2384	.2374	.2373	.7363	.7388	.7385	.4596
Distances							
Mean	2.096	2.087	1.946	1.947	2.096	2.086	2.066
σ	.021	.037	.027	.022	.030	.028	.032
Poly.Vol.	11.776	11.633	9.677	9.696	11.772	11.624	11.499
Q.E.	1.0287	1.0288	1.0102	1.0102	1.0286	1.0277	1.0155
Ang.Var.	91.0	88.1	33.3	33.5	90.6	88.4	49.2
Site Energy	-1163.	-1177.	-2540.	-2537.	-1169.	-1177.	-1230.

Table 7.4.3. Clay Mineral Tetrahedral Sites.

End-Member	Nacrite		Dickite		Kaolinite				Amesite				Lizardite
Site	T1	T2	Si1	Si2	T1	T2	T3	T4	T1	T2	T11	T22	T
C.N.	4	4	4	4	4	4	4	4	4	4	4	4	4
Occupant	Si	Si	Si	Si	Si	Si	Si	Si	$Si_{.86}Al_{.14}$	$Al_{.76}Si_{.24}$	$Al_{.73}Si_{.27}$	$Si_{.79}Al_{.21}$	Si
Point Sym.	1	1	1	1	1	1	1	1	1	1	1	1	3
Wyckoff Not.	4a	4a	4a	4a	1a	1a	1a	1a	1a	1a	1a	1a	2b
Frac.Coord.													
x	.2022	.3699	.0123	-.0013	.047	.061	.544	.579	0	-.0059	-.0117	.4900	1/3
y	.4789	-.0139	.40160	.07296	.322	.662	.831	.156	0	.3353	.0010	.1718	2/3
z	.0311	.0310	.04070	.04013	.070	.054	.048	.064	.0410	.0416	.5424	.5408	.0766
Distances													
Mean	1.617	1.620	1.612	1.612	1.619	1.641	1.659	1.569	1.639	1.729	1.725	1.649	1.638
σ	.020	.013	.009	.008	.065	.087	.082	.087	.036	.029	.010	.032	.015
Poly.Vol.	2.155	2.169	2.142	2.144	2.134	2.249	2.325	1.965	2.257	2.651	2.631	2.299	2.255
Q.E.	1.0044	1.0037	1.0024	1.0019	1.0151	1.0072	1.0076	1.0084	1.0008	1.0006	1.0005	1.0007	1.0004
Ang.Var.	17.3	15.4	10.7	8.4	40.5	24.6	20.6	26.8	0.7	1.8	1.9	1.4	1.9
Site Energy	-4225.	-4219.	-4245.	-4244.	-4343.	-4135.	-3957.	-4292.	-3934.	-2736.	-2826.	-3832.	-4112.

Chapter 8

FRAMEWORK SILICATES

Table 8.1.1. Silica Polymorph Unit Cells.

Polymorph	α-Quartz	Coesite	Stishovite	α-Cristobalite	α-Tridymite
Formula	SiO_2	SiO_2	SiO_2	SiO_2	SiO_2
Form.Wt.	60.085	60.085	60.085	60.085	60.085
Density	2.648	2.909	4.287	2.318	2.269
Mol Volume	22.688	20.657	14.017	25.925	26.478
Z	3	16	2	4	48
Cryst.Sys.	Trigonal	Monoclinic	Tetragonal	Tetragonal	Monoclinic
Cryst.Class	32	$2/m$	$4/mmm$	422	m
Space Group	$P\,3_221$	$C\,2/c$	$P\,4_2/mnm$	$P\,4_12_12$	Cc
Cell Parameters					
a	4.9134	7.1464	4.1790	4.978	18.494
b	4.9134	12.3796	4.1790	4.978	4.991
c	5.4052	7.1829	2.6651	6.948	25.832
β		120.283			117.75
Vol.	113.01	548.76	46.54	172.17	2110.2
Ref.	LePage et al. (1980)	Smyth et al.(1987)	Baur & Kahn (1971)	Peacor (1973)	Kato & Nukui (1976)

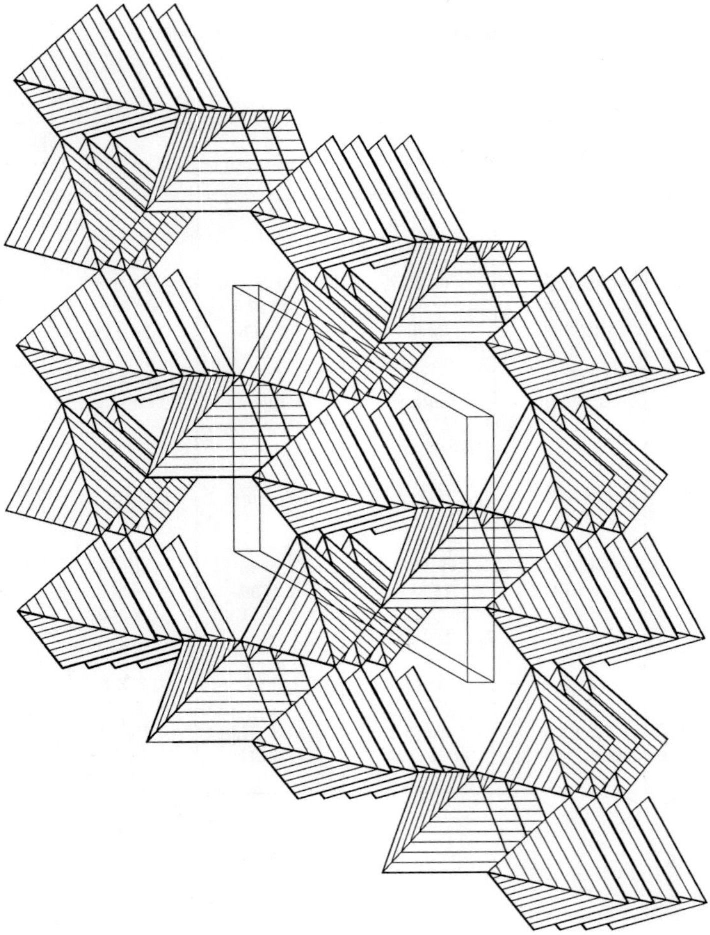

Figure 8.1a. Quartz, SiO_2, perspective c-axis projection, b vertical. The spiral framework of silicate tetrahedra can be seen looking down c. Si has point symmetry 2 and O is in the general position with point symmetry 1. There are two distinct Si–O distances in the structure.

Figure 8.1b. Coesite, SiO_2, perspective c-axis projection, a horizontal. The structure is dense and somewhat complex with interpenetrating eight-membered rings of silicate tetrahedra. There are two distinct Si and five distinct O atoms in the structure. Both silicons are in the general position. There are eight distinct $Si-O$ distances in the structure.

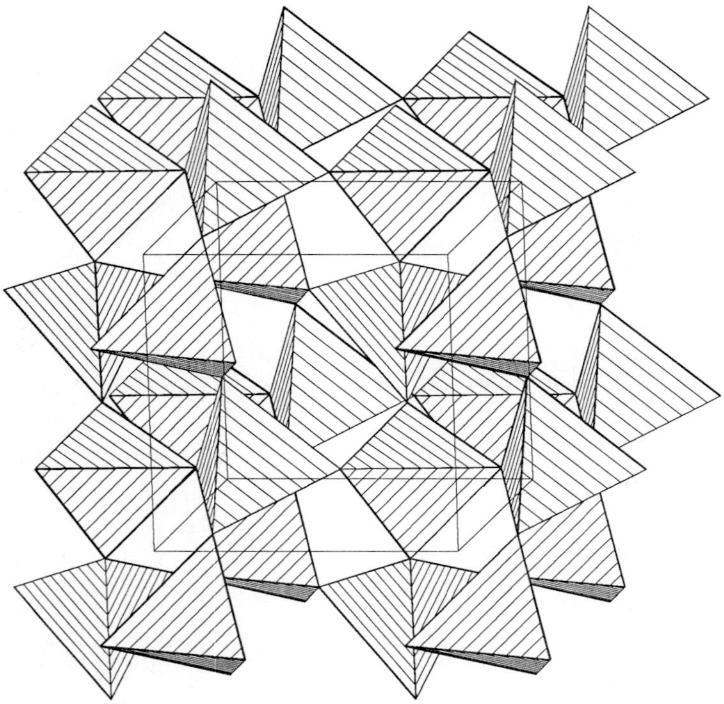

Figure 8.1c. Cristobalite, SiO_2, perspective c-axis projection. There is a single Si tetrahedron and a single oxygen in the structure. Si has point symmetry 2 and there are two distinct Si–O distances. This structure undergoes a displacive inversion to isometric symmetry at elevated temperatures.

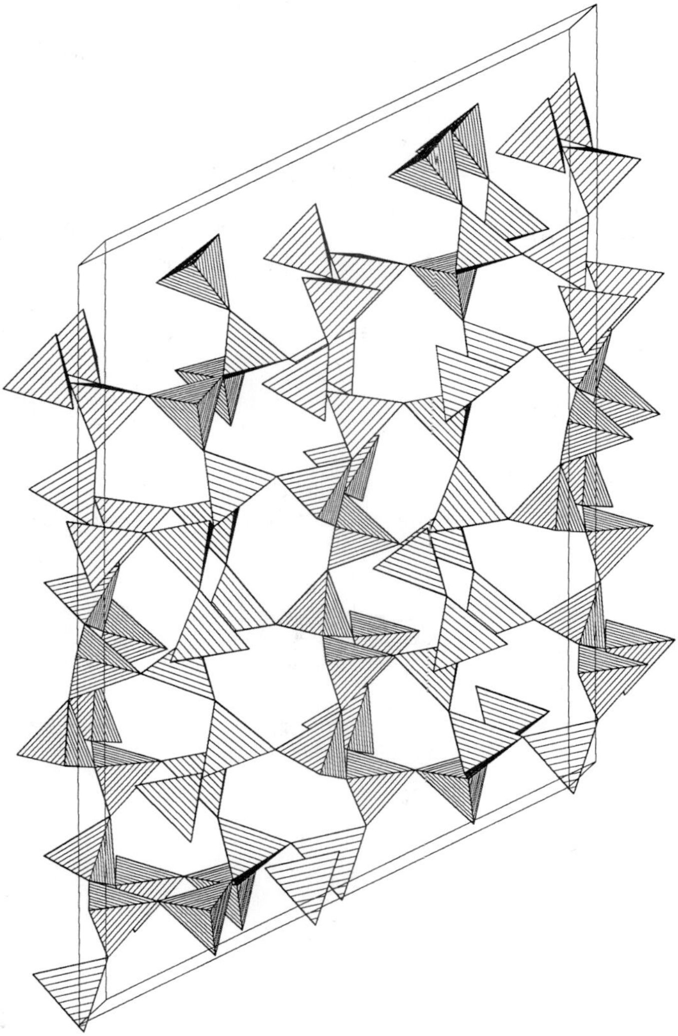

Figure 8.1d. Tridymite, SiO_2, perspective c-axis projection, b vertical. There are twelve distinct Si and 24 oxygens in the structure. This structure undergoes a displacive inversion to hexagonal symmetry at elevated temperatures, and the resulting pseudo-symmetry and complex twinning in the room temperature structure has made structure determinations difficult and uncertain.

Table 8.1.2a. Silica Polymorph Silicon Sites.

Polymorph	α-Quartz	Coesite		Stishovite	α-Cristobalite
Site	Si	Si1	Si2	Si	Si
C.N.	4	4	4	6	4
Occupant	Si	Si	Si	Si	Si
Point Sym.	2	1	1	*mmm*	2
Wyckoff Not.	3b	4e	4e	2a	4a
Frac.Coord.					
x	.46981	.14031	.50676	0	.3002
y	0	.10831	.15801	0	.3002
z	1/6	.07229	.54077	0	0
Distances					
Mean	1.609	1.611	1.614	1.775	1.606
σ	.003	.011	.011	.027	.002
Poly.Vol.	2.137	2.137	2.144	7.365	2.126
Q.E.	1.0002	1.0002	1.0002	1.0080	1.0005
Ang.Var.	0.90	1.06	0.52	27.1	2.1
Site Energy	-4459.	-4526.	-4451.	-4550.	-4445.

Table 8.1.2b. Silica Polymorph Silicon Sites (continued).

Polymorph	α-Tridymite						α-Tridymite					
Site	Si1	Si2	Si3	Si4	Si5	Si6	Si7	Si8	Si9	Si10	Si11	Si12
C.N.	4	4	4	4	4	4	4	4	4	4	4	4
Occupant	Si	Si	Si	Si	Si	Si	Si	Si	Si	Si	Si	Si
Point Sym.	1	1	1	1	1	1	1	1	1	1	1	1
Wyckoff Not.	8f	8f	8f	8f	8f	8f	8f	8f	8f	8f	8f	8f
Frac.Coord.												
x	.0200	.2067	.0349	.2228	.0538	.2427	.2680	.4551	.2929	.4837	.2991	.4834
y	.1896	.2899	.7046	.8046	.1992	.2059	.7848	.7040	.3006	.1994	.7130	.8083
z	.03800	.09691	.20212	.26187	.37631	.43476	.06278	.12843	.23294	.30132	.39594	.46057
Distances												
Mean	1.605	1.598	1.600	1.597	1.599	1.588	1.604	1.590	1.581	1.588	1.597	1.596
σ	.024	.028	.010	.024	.022	.030	.019	.016	.014	.018	.026	.011
Poly.Vol.	2.119	2.091	2.099	2.087	2.097	2.052	2.115	2.062	2.026	2.054	2.088	2.086
Q.E.	1.0005	1.0007	1.0003	1.0006	1.0006	1.0007	1.0012	1.0002	1.0005	1.0006	1.0009	1.0003
Ang.Var.	1.46	1.82	1.25	1.90	1.42	1.10	4.33	0.65	2.09	2.38	1.96	1.06
Site Energy	-4426.	-4445.	-4469.	-4493.	-4463.	-4507.	-4438.	-4488.	-4554.	-4525.	-4464.	-4450.

Table 8.2.1. Alkali Feldspar Unit Cells.

End-Member	Sanidine	Orthoclase	Microcline	Rb-Feldspar	High Albite	Low Albite
Formula	$KAlSi_3O_8$	$KAlSi_3O_8$	$KAlSi_3O_8$	$RbAlSi_3O_8$	$NaAlSi_3O_8$	$NaAlSi_3O_8$
Form.Wt.	278.337	278.337	278.337	324.705	262.225	262.225
Density	2.578	2.571	2.567	2.930	2.610	2.621
Mol Volume	107.958	108.283	108.425	110.825	100.452	100.054
Z	4	4	4	4	4	4
Cryst.Sys.	Monoclinic	Monoclinic	Triclinic	Monoclinic	Triclinic	Triclinic
Cryst.Class	$2/m$	$2/m$	$\bar{1}$	$2/m$	$\bar{1}$	$\bar{1}$
Space Group	$C2/m$	$C2/m$	$C\bar{1}$	$C2/m$	$C\bar{1}$	$C\bar{1}$
Cell Parameters						
a	8.539	8.561	8.560	8.820	8.161	8.142
b	13.015	12.996	12.964	12.992	12.875	12.785
c	7.179	7.192	7.215	7.161	7.110	7.159
α			90.65		93.53	94.19
β	115.99	116.01	115.83	116.24	116.46	116.61
γ			87.70		90.24	87.68
Vol.	717.15	719.13	720.07	736.01	667.12	664.48
Ref.	Phillips & Ribbe (1973)	Colville & Ribbe (1968)	Brown & Bailey(1964)	Gasperin (1971)	Winter et al. (1979)	Harlow & Brown (1980)

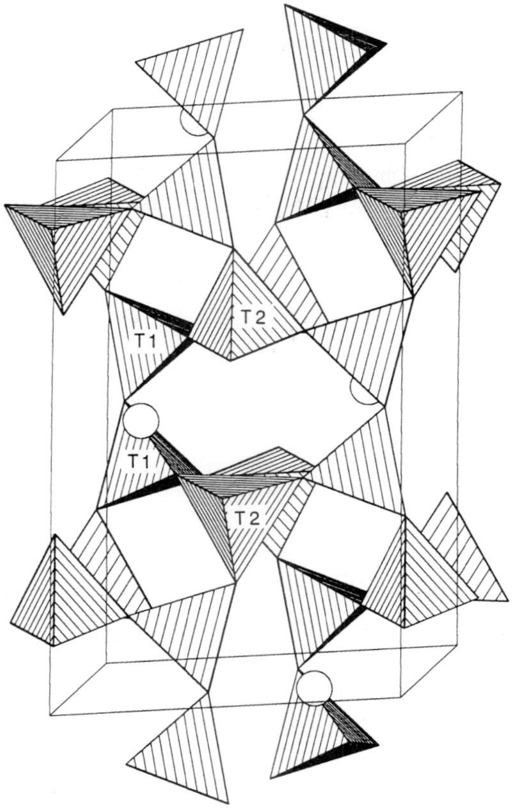

Figure 8.2. Albite, $NaAlSi_3O_8$, perspective c-axis projection, b vertical. This is an illustration of the basic framework common to all feldspars. The topologic symmetry of the framework is monoclinic, $C2/m$, and sanidine and disordered albite at high temperatures have this maximum symmetry with two distinct Si/Al tetrahedra, T1 and T2. Si/Al ordering and collapse of the framework around the smaller Na atoms (circles) can both result in lowering of the symmetry to triclinic, $C\bar{1}$. The triclinic structure has four distinct tetrahedral sites, T10, T1m, T20, and T2m. Ordering of Al and Si may then result in all of the Al residing in T10. Additional Al cannot be placed in the ordered structure without resulting in Al-O-Al bonds which are energetically unfavorable, so the alkaline earth feldspars have a more complex ordering scheme with a doubled (14Å) c-axis. Anorthite at room temperature has sixteen distinct tetrahedral sites, half of which contain Al and half Si in a strongly ordered arrangement.

Table 8.2.2. Alkali Feldspar Alkali Sites.

End-Member	Sanidine	Orthoclase	Microcline	Rb-feldspar	High Albite	Low Albite
C.N.	10	10	10	10	6	6
Occupant	K	K	K	Rb	Na	Na
Point Sym.	*m*	*m*	1	*m*	1	1
Wyckoff Not.	2m	2m	2i	2m	2i	2i
Frac.Coord.						
x	.2839	.2838	.2827	.2958	.2737	.26849
y	0	0	-.0072	0	.0075	.98570
z	.1366	.1373	.1365	.1468	.1332	.14672
Distances						
Mean	3.002	3.014	3.013	3.100	2.597	2.546
σ	.187	.186	.212	.085	.198	.215
Poly.Vol.	62.86	63.52	51.27	67.48	17.92	17.19
Q.E.					1.1990	1.1860
Ang.Var.					553.	537.
Site Energy	-276.	-271.	-273.	-254.	-310.	-333.

Table 8.2.3a. Alkali Feldspar Tetrahedral Sites.

End-Member	Sanidine		Orthoclase		Rb-Feldspar		Microcline			
Site	T_1	T_2	T_1	T_2	T_1	T_2	$T_1(0)$	$T_1(m)$	$T_2(0)$	$T_2(m)$
C.N.	4	4	4	4	4	4	4	4	4	4
Occupant	$Al_{.25}Si_{.75}$	$Al_{.25}Si_{.75}$	$Al_{.5}Si_{.5}$	Si	$Al_{.25}Si_{.75}$	$Al_{.25}Si_{.75}$	Al	Si	Si	Si
Point Sym.	1	1	1	1	1	1	1	1	1	1
Wyckoff Not.	2o	2o	2o	2o	2o	2o	2i	2i	2i	2i
Frac.Coord.										
x	.0094	.7079	.0095	.7089	.0104	.7227	.0104	.0097	.7110	.7059
y	.1843	.1177	.1844	.1178	.1904	.1195	.1875	.8198	.1202	.8856
z	.2243	.3444	.2239	.3443	.2227	.3440	.2169	.2327	.3399	.3507
Distances										
Mean	1.649	1.637	1.655	1.629	1.636	1.634	1.714	1.614	1.610	1.611
σ	.008	.004	.011	.009	.005	.008	.003	.017	.026	.024
Poly.Vol.	2.293	2.246	2.321	2.211	2.241	2.234	2.694	2.155	2.136	2.137
Q.E.	1.0026	1.0019	1.0022	1.0017	1.0016	1.0013	1.0032	1.0014	1.0025	1.0032
Ang.Var.	10.08	7.75	8.72	7.31	6.09	5.12	12.44	4.95	10.04	13.55
Site Energy	-3863.	-3902.	-3469.	-4337.	-3912.	-3913.	-2576.	-4367.	-4368.	-4431.
Model Charge	3.75	3.75	3.5	4.0	3.75	3.75	3.0	4.0	4.0	4.0

Table 8.2.3b. Alkali Feldspar Tetrahedral Sites (continued).

End-Member	High Albite				Low Albite			
Site	$T_1(0)$	$T_1(m)$	$T_2(0)$	$T_2(m)$	$T_1(0)$	$T_1(m)$	$T_2(0)$	$T_2(m)$
C.N.	4	4	4	4	4	4	4	4
Occupant	$Al_{.25}Si_{.75}$	$Al_{.25}Si_{.75}$	$Al_{.25}Si_{.75}$	$Al_{.25}Si_{.75}$	Al	Si	Si	Si
Point Sym.	1	1	1	1	1	1	1	1
Wyckoff Not.	2i	2i	2i	2i	2i	2i	2i	2i
Frac.Coord.								
x	.0090	.0048	.6904	.6849	.00901	.00386	.69201	.68152
y	.1649	.8146	.1080	.8726	.16862	.82062	.11036	.88195
z	.2147	.2289	.3202	.3537	.20806	.23728	.31508	.36078
Distances								
Mean	1.649	1.642	1.641	1.643	1.750	1.624	1.614	1.616
σ	.005	.012	.011	.010	.019	.024	.016	.022
Poly.Vol.	2.287	2.267	2.263	2.273	2.700	2.185	2.151	2.158
Q.E.	1.0040	1.0019	1.0017	1.0007	1.0128	1.0044	1.0022	1.0021
Ang.Var.	15.97	7.42	7.02	3.03	51.26	15.78	9.57	8.99
Site Energy	-3839.	-3915.	-3911.	-3880.	-2530.	-4363.	-4375.	-4393.
Model Charge	3.75	3.75	3.75	3.75	3.0	4.0	4.0	4.0

Table 8.3.1. Alkaline Earth Feldspar Unit Cells.

End-Member	Anorthite	Celsian
Formula	$CaAl_2Si_2O_8$	$BaAl_2Si_2O_8$
Form.Wt.	278.210	375.470
Density	2.765	3.400
Mol Volume	100.610	110.440
Z	8	8
Cryst.Sys.	Triclinic	Monoclinic
Cryst.Class	$\bar{1}$	$2/m$
Space Group	$P\bar{1}$	$I2/m$
Cell Parameters		
a	8.173	8.627
b	12.869	13.045
c	14.165	14.408
α	93.113	90
β	115.913	115.22
γ	91.261	90
Vol.	1336.35	1466.90
Ref.	Wainwright & Starkey (1971)	Newnham & Megaw (1960)

Table 8.3.2. Alkaline Earth Feldspar Ca and Ba Sites.

End-Member	Anorthite				Celsian
Site	A(000)	A(z00)	A(0i0)	A(zi0)	Ba
C.N.	5	6	6	6	9
Occupant	Ca	Ca	Ca	Ca	Ba
Point Sym.	1	1	1	1	m
Wyckoff Not.	2b	2b	2b	2b	2i
Frac.Coord.					
x	.2651	.2692	.7737	.7634	.2826
y	.9864	.0312	.5359	.5052	0
z	.0867	.5435	.5412	.0747	.0653
Distances					
Mean	2.485	2.516	2.529	2.572	2.921
σ	.100	.123	.150	.173	.141
Poly.Vol.	7.861	13.168	12.549	16.204	41.165
Q.E.	-	1.378	1.439	1.256	-
Ang.Var.	-	963.9	41.4	531.8	-
Site Energy	-959.	-957.	-953.	-975.	-809.

Table 8.3.3. Alkaline Earth Feldspar Aluminum Sites.

End-Member	Anorthite				Anorthite				Celsian	
Site	$T_1(0z00)$	$T_1(0zi0)$	$T_1(m000)$	$T_1(m0i0)$	$T_2(0000)$	$T_2(00i0)$	$T_2(mz00)$	$T_2(mzi0)$	$T_1(0z00)$	$T_2(0000)$
C.N.	4	4	4	4	4	4	4	4	4	4
Occupant	Al	Al	Al	Al	Al	Al	Al	Al	Al	Al
Point Sym.	1	1	1	1	1	1	1	1	1	1
Wyckoff Not.	2*b*	2*b*	2*b*	2*b*	2*b*	2*b*	2*b*	2*b*	2*b*	2*b*
Frac.Coord.										
x	.0066	.4984	.9912	.5073	.6845	.1907	.6809	.1852	.0073	.7058
y	.1610	.6658	.8152	.3145	.1130	.6110	.8719	.3775	.1832	.1205
z	.6112	.1128	.1176	.6212	.1519	.6674	.6725	.1816	.6142	.1733
Distances										
Mean	1.747	1.755	1.749	1.744	1.741	1.750	1.744	1.745	1.719	1.712
σ	.029	.020	.035	.031	.031	.017	.019	.028	.014	.015
Poly.Vol.	2.667	2.667	2.718	2.690	2.694	2.690	2.699	2.716	2.569	2.547
Q.E.	1.0171	1.0269	1.0075	1.0079	1.0039	1.0146	1.0057	1.0024	1.0099	1.0069
Ang.Var.	64.56	99.96	29.84	30.27	14.98	54.13	22.06	9.04	38.37	24.33
Site Energy	-2484.	-2488.	-2563.	-2562.	-2586.	-2501.	-2509.	-2557.	-2456.	-2525.

Table 8.3.4. Alkaline Earth Feldspar Silicon Sites.

End-Member	Anorthite				Anorthite				Celsian	
Site	$T_1(0000)$	$T_1(00i0)$	$T_1(mz00)$	$T_1(mzi0)$	$T_2(0z00)$	$T_2(0zi0)$	$T_2(m000)$	$T_2(m0i0)$	$T_1(0000)$	$T_2(0z00)$
C.N.	4	4	4	4	4	4	4	4	4	4
Occupant	Si	Si	Si	Si	Si	Si	Si	Si	Si	Si
Point Sym.	1	1	1	1	1	1	1	1	1	1
Wyckoff Not.	$2b$	$2b$	$2b$	$2b$	$2b$	$2b$	$2b$	$2b$	$2b$	$2b$
Frac.Coord.										
x	.0092	.5062	.0061	.5041	.6841	.1713	.6742	.1762	.0091	.7004
y	.1592	.6560	.8154	.3204	.1034	.6067	.8829	.3789	.1828	.1165
z	.1044	.6042	.6135	.1099	.6646	.1495	.1876	.6734	.1096	.6735
Distances										
Mean	1.615	1.612	1.613	1.613	1.616	1.607	1.615	1.615	1.637	1.635
σ	.026	.020	.031	.027	.014	.024	.028	.022	.008	.012
Poly.Vol.	2.125	2.125	2.133	2.137	2.146	2.117	2.153	2.159	2.226	2.228
Q.E.	1.0121	1.0083	1.0061	1.0048	1.0065	1.0047	1.0027	1.0015	1.0079	1.0043
Ang.Var.	47.86	32.90	23.23	19.42	25.61	18.35	10.84	6.39	30.48	15.99
Site Energy	-4412.	-4402.	-4467.	-4500.	-4431.	-4532.	-4472.	-4445.	-4395.	-4410.

Table 8.4.1. Feldspathoid Group Unit Cells.

End-Member	Leucite	Kalsilite	Nepheline
Formula	$KAlSi_2O_6$	$KAlSiO_4$	$KNa_3Al_4Si_4O_{16}$
Form.Wt.	218.252	158.167	584.332
Density	2.461	2.621	2.679
Mol Volume	88.69	60.34	218.09
Z	16	2	2
Cryst.Sys.	Tetragonal	Hexagonal	Hexagonal
Cryst.Class	$4/m$	6	6
Space Group	$I\,4_1/a$	$P\,6_3$	$P\,6_3$
Cell Parameters			
a	13.09	5.16	9.993
c	13.75	8.69	8.374
Vol.	2356.	200.4	724.19
Ref.	Mazzi et al.(1976)	Perrotta & Smith (1967)	Foreman & Peacor(1970)

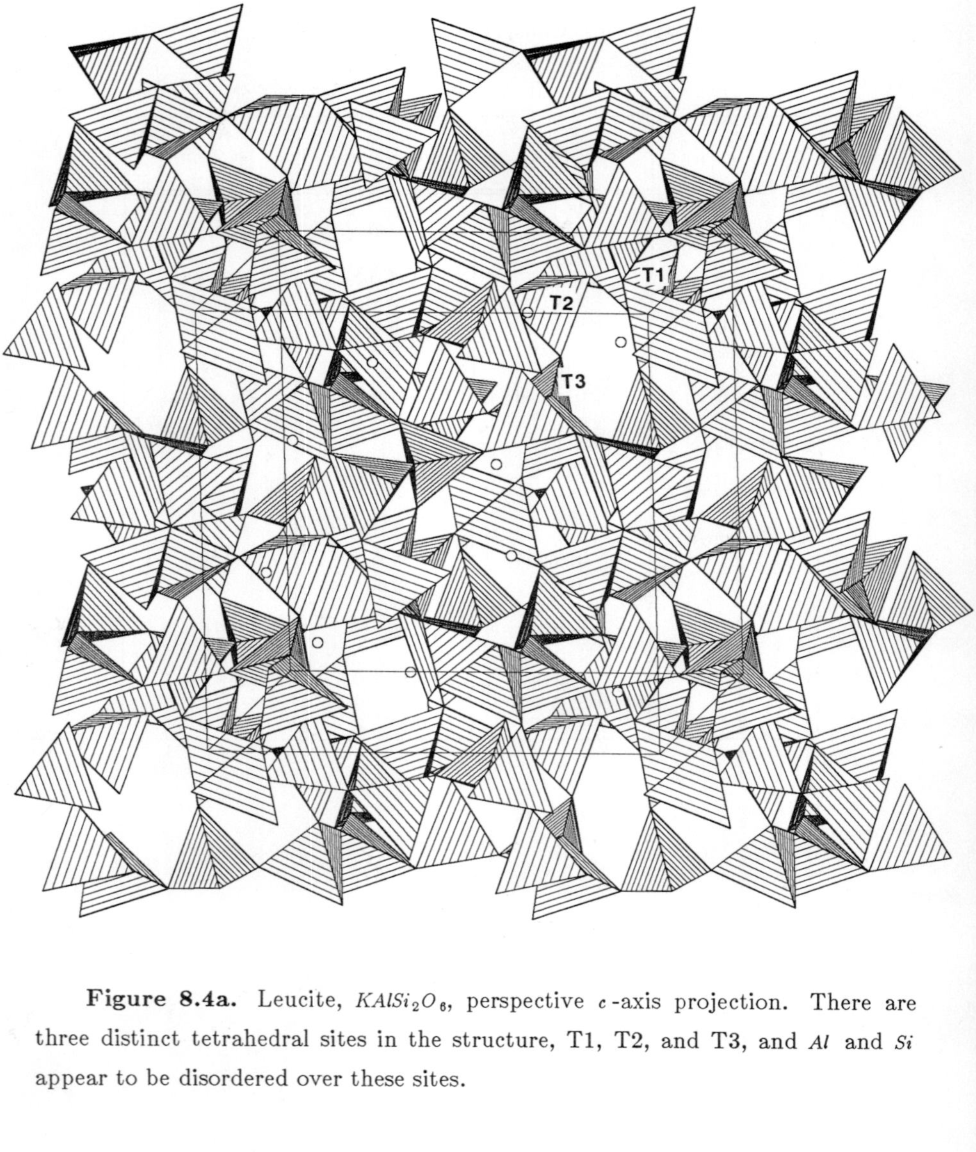

Figure 8.4a. Leucite, $KAlSi_2O_6$, perspective *c*-axis projection. There are three distinct tetrahedral sites in the structure, T1, T2, and T3, and *Al* and *Si* appear to be disordered over these sites.

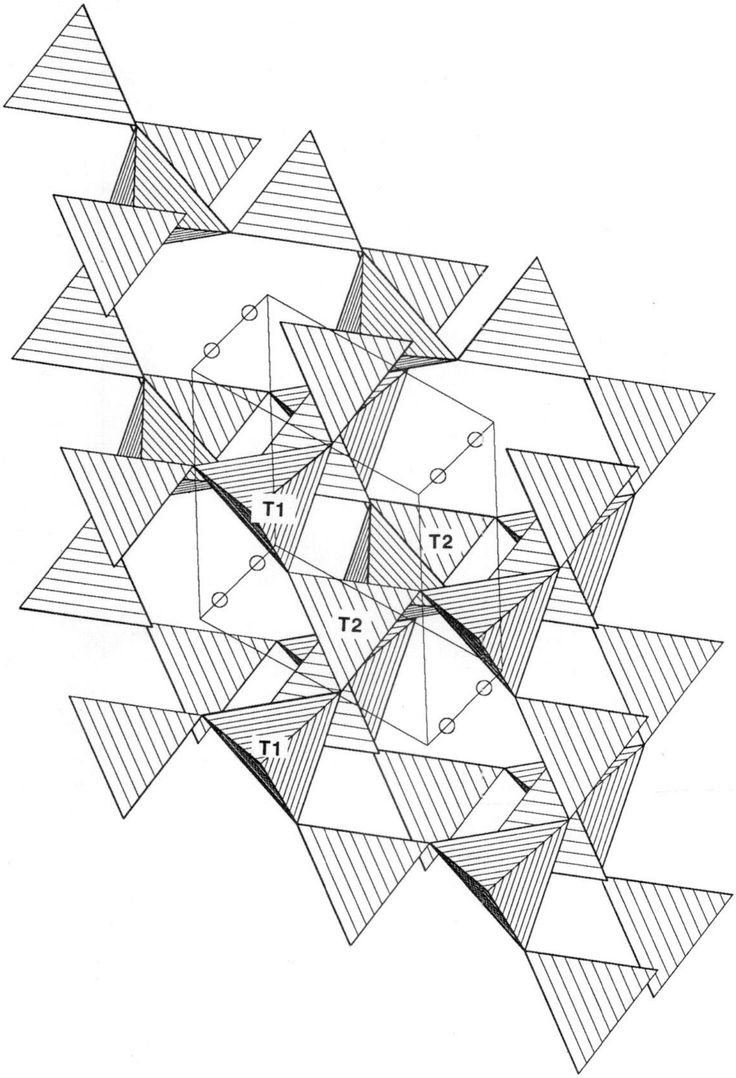

Figure 8.4b. Kalsilite, $KAlSiO_4$, perspective *c*-axis projection. There are two distinct tetrahedral sites in the structure, T1 and T2, and *Al* and *Si* appear to be ordered with *Al* in T1. Nepheline is an ordered derivative of this structure with a doubled *a*-axis and four distinct tetrahedra.

Table 8.4.2. Feldspathoid Group Alkali Sites.

Member	Leucite	Kalsilite	Nepheline	
Site	K	K	K	Na
C.N.	6	9	9	8
Occupant	K	K	K	Na
Point Sym.	1	3	3	1
Wyckoff Not.	16f	2a	2a	6c
Frac.Coord.				
x	.3660	0	0	.4425
y	.3645	0	0	.9968
z	.1147	.2411	.9883	.9930
Distances				
Mean	3.015	2.968	3.016	2.620
σ	.067	.024	.030	.119
Poly.Vol.	23.95	46.69	49.93	26.74
Q.E.	1.3256			
Ang.Var.	951.			
Site Energy	-237.	-226.	-190.	-278.

Table 8.4.3. Feldspathoid Group Tetrahedral Sites.

Member	Leucite			Kalsilite		Nepheline			
Site	T1	T2	T3	T1	T2	T1	T2	T3	T4
C.N.	4	4	4	4	4	4	4	4	4
Occupant	$Si_{.67}Al_{.33}$	$Si_{.67}Al_{.33}$	$Si_{.67}Al_{.33}$	Al	Si	Al	Si	Si	Al
Point Sym.	1	1	1	3	3	3	3	1	1
Wyckoff Not.	16f	16f	16f	2b	2b	2b	2b	6c	6c
Frac.Coord.									
x	.0579	.1676	.3924	1/3	1/3	2/3	2/3	.3343	.3322
y	.3964	.6115	.6418	2/3	2/3	1/3	1/3	.0934	.0930
z	.1666	.1283	.0860	.0461	.4281	.1879	.8008	.3095	.6831
Distances									
Mean	1.644	1.657	1.657	1.744	1.609	1.713	1.612	1.616	1.734
σ	.007	.008	.014	.030	.027	.005	.008	.028	.021
Poly.Vol.	2.271	2.329	2.331	2.698	2.113	2.524	2.109	2.162	2.671
Q.E.	1.0022	1.0014	1.0006	1.0061	1.0079	1.0147	1.0131	1.0012	1.0017
Ang.Var.	8.73	5.39	2.28	23.95	27.58	54.62	48.96	3.72	5.80
Site Energy	-3806.	-3717.	-3668.	-2563.	-4505.	-2658.	-4510.	-4455.	-2558.
Model Charge	3.67	3.67	3.67	3.0	4.0	3.00	4.00	4.00	3.00

Table 8.5.1. Beryl and Cordierite Unit Cells.

End-Member	Beryl	Cordierite
Formula	$Be_3Al_2Si_6O_{18}$	$Mg_2Si_5Al_4O_{18}$
Form.Wt.	537.505	584.97
Density	2.645	2.499
Mol Volume	203.24	234.11
Z	2	4
Cryst.Sys.	Hexagonal	Orthorhombic
Cryst.Class	$6/mmm$	mmm
Space Group	$P6/mcc$	$Cccm$
Cell Parameters		
a	9.2086	17.079
b	9.2086	9.730
c	9.1900	9.356
Vol.	674.89	1554.77
Ref.	Morosin (1972)	Cohen et al. (1977)

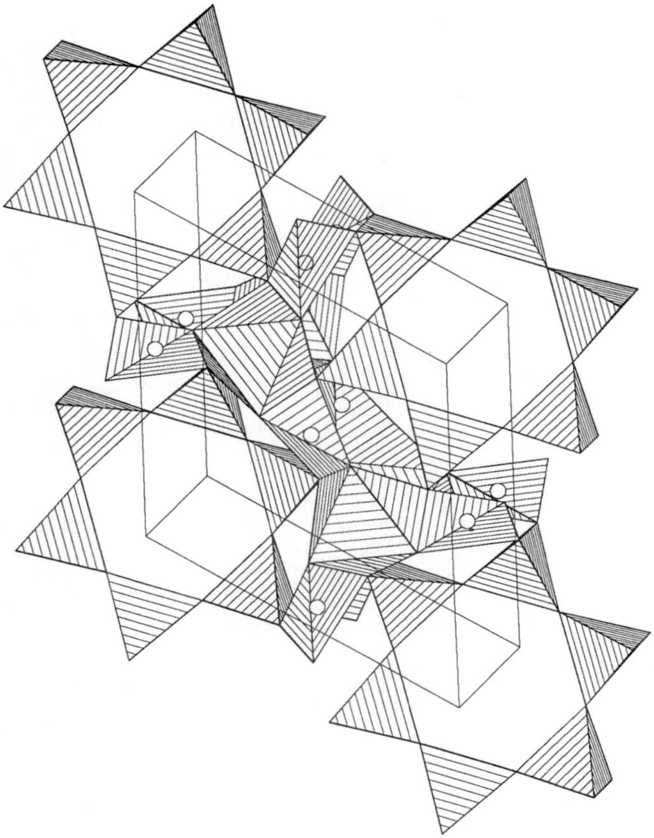

Figure 8.5. Beryl, $Be_3Al_2Si_6O_{18}$, perspective c-axis projection. Beryl and cordierite have very similar structures, although an interstitial alkali site is occupied by Na or K in cordierite but may be vacant in beryl. In some varieties of beryl, Cs or Rb may partially occupy this site. In beryl, there is a single Be tetrahedron (circle) with point symmetry 222. In cordierite this site is occupied by Al. In beryl, there is a single Al octahedron with point symmetry 32, which in cordierite is occupied by Mg. Beryl also has a single Si tetrahedron with point symmetry m which forms an Si_6O_{18} ring. The ring tetrahedra are partially occupied by Al in cordierite.

Table 8.5.2. Beryl Cation Sites.

End-Member	Beryl		
Site	Be	Al	Si
C.N.	4	6	4
Occupant	Be	Al	Si
Point Sym.	222	32	*m*
Wyckoff Not.	6f	4c	12l
Frac.Coord.			
x	1/2	2/3	.38749
y	0	1/3	.11587
z	1/4	1/4	0
Distances			
Mean	1.654	1.906	1.608
σ	.001	.001	.015
Poly.Vol.	2.027	8.939	2.132
Q.E.	1.0950	1.0218	1.0006
Ang.Var.	331.7	75.53	1.69
Site Energy	-1348.	-2086.	-4474.

Table 8.5.3. Cordierite Octahedral and Alkali Sites.

Site	M	Alkali
C.N.	6	8
Occupant	Mg	Na,K
Point Sym.	2	2/m
Wyckoff Not.	8g	4c
Frac.Coord.		
x	.1625	0
y	1/2	0
z	1/4	0
Distances		
	O1-1(2) 2.100	O2-1(2) 2.750
	O1-6(2) 2.114	O2-6(2) 2.519
	O1-3(2) 2.116	Ow (4) 2.384
Mean	2.110	2.509
σ	.008	.160
Poly.Vol.	11.798	23.256
Q.E.	1.0406	
Ang.Var.	128.4	
Site Energy	-1119.	0.0
Model Charge	2.0	0.0

Table 8.5.4. Cordierite Tetrahedral Sites.

Site	T1-1	T2-6	T1-6	T2-1	T2-3
C.N.	4	4	4	4	4
Occupant	Al	Al	Si	Si	Si
Point Sym.	2	m	222	m	m
Wyckoff Not.	8k	8l	4b	8l	8l
Frac.Coord.					
x	1/4	.0508	0	.1926	-.1352
y	1/4	.3079	1/2	.0778	.2375
z	.2502	0	1/4	0	0
Distances					
Mean	1.758	1.742	1.626	1.614	1.617
σ	.001	.036	.000	.026	.027
Poly.Vol.	2.581	2.705	2.133	2.155	2.166
Q.E.	1.0530	1.0025	1.0226	1.0010	1.0015
Ang.Var.	195.7	10.6	88.6	4.4	6.3
Site Energy	-2419.	-2596.	-4327.	-4450.	-4429.
Model Charge	3.0	3.0	4.0	4.0	4.0

Table 8.6.1. Scapolite Group Unit Cells.

End-Member	Marialite	Meionite
Formula(Ideal)	$Na_4Al_3Si_9O_{24}Cl$	$Ca_4Al_6Si_6O_{24}CO_3$
Formula(obs.)	$Na_{2.9}Ca_{.8}K_{.2}$ $Al_{3.7}Si_{8.3}O_{24.2}$ $Cl_{.7}(CO_3)_{.2}$	$Ca_{3.7}Na_{.2}K_{.1}$ $Al_{5.7}Si_{6.2}O_{24.2}$ $Cl_{.03}(CO_3)_{.95}(SO_4)_{.03}$
Form.Wt.(obs.)	863.5	932.9
Density	2.599	2.757
Mol Volume	332.26	338.40
Z	2	2
Cryst.Sys.	Tetragonal	Tetragonal
Cryst.Class	$4/m$	$4/m$
Space Group	$P\,4_2/n$	$P\,4_2/n$
Cell Parameters		
a	12.059	12.194
c	7.587	7.557
Vol.	1103.3	1123.7
Ref.	Lin & Burley (1973a)	Lin & Burley (1973b)

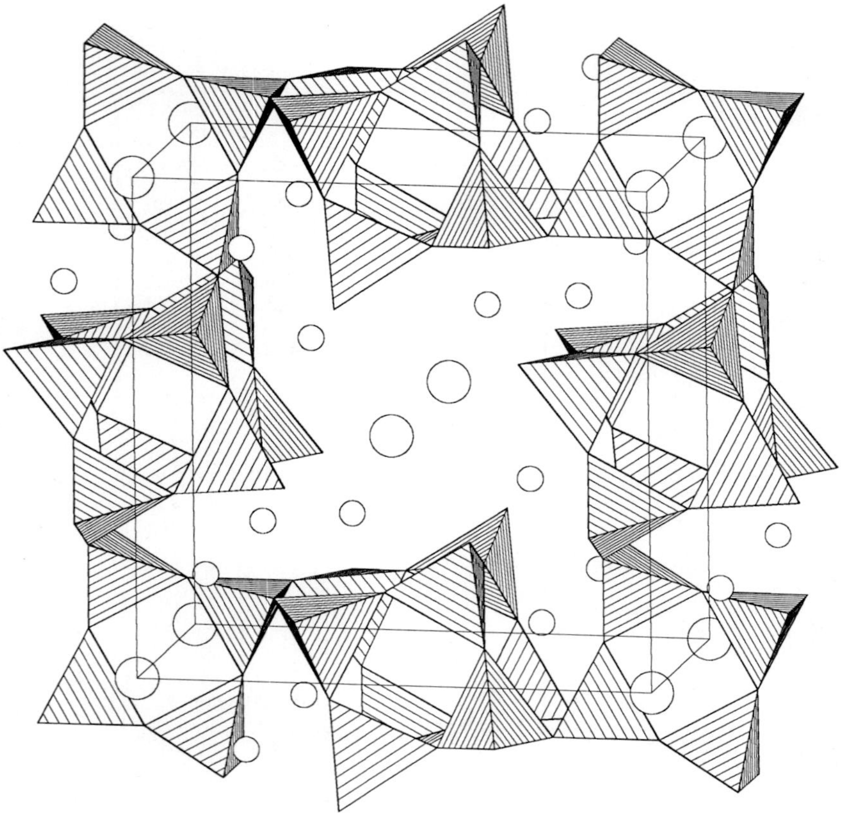

Figure 8.6. Marialite, $Na_4Al_3Si_9O_{24}Cl$, Perspective c-axis projection. There are three distinct tetrahedral sites which may contain Si or Al in varying degrees of order. Na (Small circles) and Cl (large circles) occupy the cavity sites.

Table 8.6.2. Scapolite Group M Sites.

End-Member	Marialite	Meionite
C.N.	9	9
Occupant	$Na_{.71}Ca_{.21}K_{.06}$	$Na_{.05}Ca_{.92}K_{.03}$
Point Sym.	1	1
Wyckoff Not.	8g	8g
Frac.Coord.		
x	.3659	.3565
y	.2884	.2801
z	.5064	.5001
Distances		
O2	2.350	2.354
O3	2.517	2.493
O4	2.501	2.491
O5	2.869	2.870
O5'	2.817	2.679
O6	2.882	2.655
O6'	2.946	2.894
O7	2.341	2.564
O7'	2.166	2.402
Mean	2.599	2.600
σ	.286	.192
Poly.Vol.	32.548	32.597
Site Energy	-416.	-909.
Model Charge	1.21	1.92

Table 8.6.3. Scapolite Group Tetrahedral Sites.

End-Member	Marialite			Meionite		
C.N.	4	4	4	4	4	4
Occupant	Si	$Al_{.59}Si_{.41}$	$Al_{.29}Si_{.71}$	$Al_{.31}Si_{.69}$	$Al_{.51}Si_{.49}$	$Al_{.49}Si_{.51}$
Point Sym.	1	1	1	1	1	1
Wyckoff Not.	8g	8g	8g	8g	8g	8g
Frac.Coord.						
x	.3390	.6621	.4145	.3395	.6599	.4130
y	.4100	.9152	.8368	.4074	.9129	.8401
z	.9989	.7956	.7085	.0000	.7930	.7072
Distances						
1	1.594	1.688	1.646	1.634	1.670	1.679
2	1.612	1.690	1.639	1.674	1.681	1.680
3	1.607	1.693	1.673	1.628	1.681	1.682
4	1.619	1.699	1.644	1.674	1.690	1.673
Mean	1.608	1.693	1.651	1.653	1.681	1.679
σ	.011	.005	.015	.025	.008	.004
Poly.Vol.	2.133	2.474	2.292	2.309	2.414	2.403
Q.E.	1.0003	1.0041	1.0043	1.0022	1.0061	1.0063
Ang.Var.	1.60	16.49	17.46	9.62	23.88	24.74
Site Energy	-4414.	-3137.	-3718.	-3802.	-3263.	-3301.
Model Charge	4.0	3.41	3.71	3.69	3.49	3.51

Table 8.7.1a. Zeolite Group Unit Cells.

End-Member	Analcime	Chabazite	K-Mordenite
Formula	$Na_{15.15}Ca_{0.6}$ $Al_{16.33}Si_{31.65}O_{96}$ $16H_2O$	$Ca_{1.4}Sr_{.3}$ $Al_{3.8}Si_{8.3}O_{24}$ $13H_2O$	K_8 $Al_8Si_{40}O_{96}$ $24H_2O$
Form.Wt.	3526.1	1036.23	3620.4
Density	2.264	2.075	2.132
Mol Volume	1557.4	499.35	1698.0
Z	1	1	1
Cryst.Sys.	Tetragonal	Trigonal	Orthorhombic
Cryst.Class	$4/mmm$	$\bar{3}m$	mmm
Space Group	$I4_1/acd$	$R\bar{3}m$	$Cmcm$
Cell Parameters			
a	13.721	9.421	18.167
b	13.721	9.421	20.611
c	13.735	9.421	7.529
α		94.20	
Vol.	2585.8	829.08	2819.2
Ref.	Mazzi & Galli (1978)	Calligaris et al.(1982)	Mortier et al. (1978)

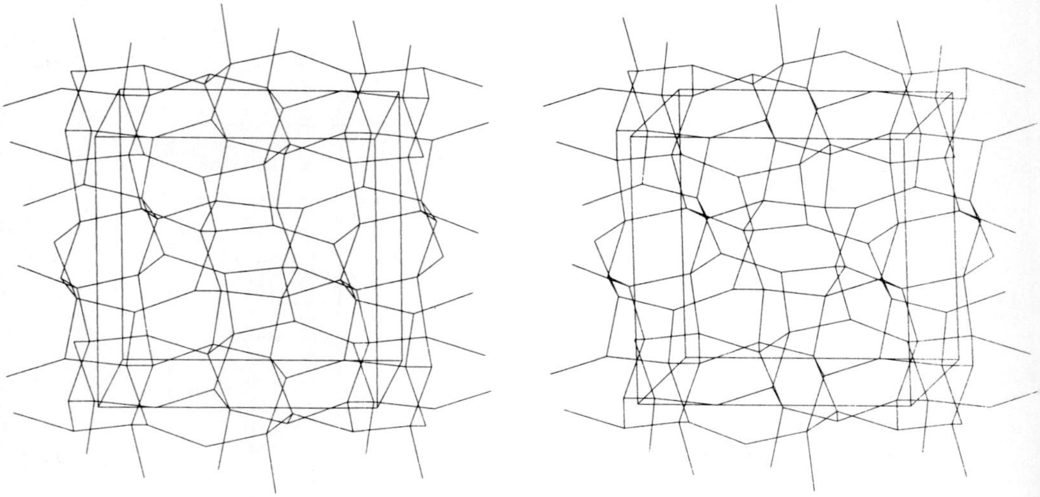

Figure 8.7a. Analcime, $NaAlSi_2O_6 H_2O$, stereoscopic c-axis projection. The framework is shown schematically.

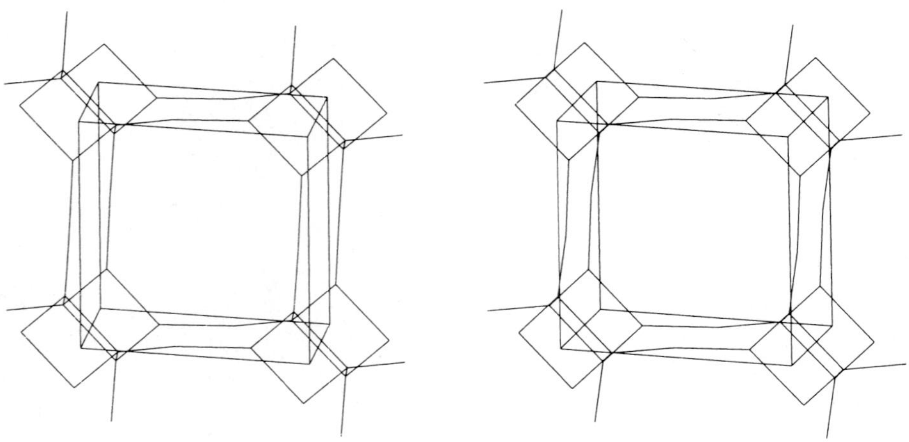

Figure 8.7b. Chabazite, $Ca_2Al_4Si_8O_{24}13H_2O$, stereoscopic c-axis projection. The framework is shown schematically.

Table 8.7.1b. Zeolite Group Unit Cells (continued).

End-Member	Clinoptilolite	Heulandite	K-Heulandite
Formula	$Ca_{1.16}Na_{1.8}$ $K_{1.68}Mg_{0.3}$ $Al_{6.3}Si_{29.8}O_{72}$ $24H_2O$	$Ca_{3.6}Na_{0.3}$ $K_{1.0}Sr_{0.3}$ $Al_{9.5}Si_{26.6}O_{72}$ $25.5H_2O$	$Na_{.1}K_{8.6}$ $Ba_{.04}$ $Al_{9.3}Si_{26.8}O_{72}$ $19.6H_3O$
Form.Wt.	2750.6	2831.3	2852.7
Density	2.176	2.236	2.221
Mol Volume	1264.1	1266.4	1284.3
Z	1	1	1
Cryst.Sys.	Monoclinic	Monoclinic	Monoclinic
Cryst.Class	$2/m$	$2/m$	$2/m$
Space Group	$C2/m$	$C2/m$	$C2/m$
Cell Parameters			
a	17.662	17.715	17.767
b	17.911	17.831	17.958
c	7.407	7.430	7.431
β	116.40	116.38	115.93
Vol.	2098.8	2102.6	2132.2
Ref.	Koyama & Takeuchi (1977)	Alberti & Vezzalini (1983)	Galli et al. (1983)

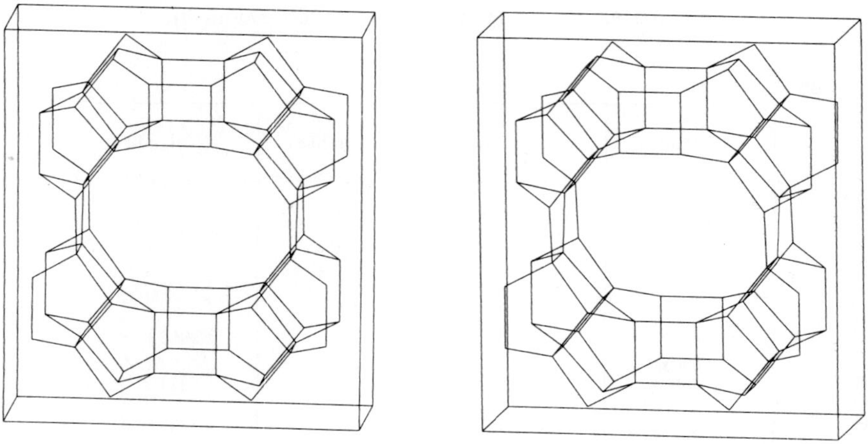

Figure 8.7c. Mordenite, $(Na,K)_8Al_8Si_{40}O_{96}24H_2O$, stereoscopic c-axis projection. The framework is shown schematically.

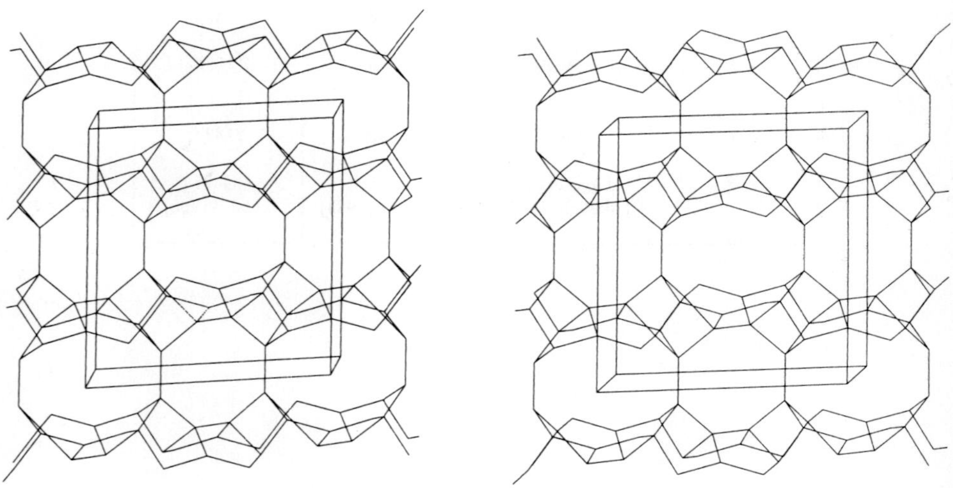

Figure 8.7d. Clinoptilolite, $(CaNa)Al_2Si_7O_{18}6H_2O$, stereoscopic c-axis projection. The framework is shown schematically. Heulandite has a similar framework.

Table 8.7.1c. Zeolite Group Unit Cells (continued).

End-Member	Thomsonite	Harmotome	Phillipsite	Laumontite
Formula	$NaCa_2$ $Al_5Si_5O_{20}$ $6H_2O$	$Ba_2Ca_{0.5}$ $Al_5Si_{11}O_{32}$ $12H_2O$	$K_2Ca_{1.5}Na_{0.4}$ $Al_5Si_{10}O_{32}$ $12H_2O$	Ca $Al_2Si_4O_{12}$ $4H_3O$
Form.Wt.	806.6	1466.7	1291.5	470.44
Density	2.373	2.443	2.120	2.315
Mol Volume	339.9	600.5	609.1	203.2
Z	4	1	1	4
Cryst.Sys.	Orthorhombic	Monoclinic	Monoclinic	Monoclinic
Cryst.Class	*mmm*	$2/m$	$2/m$	*m*
Space Group	*Pncn*	$P\,2_1/m$	$P\,2_1/m$	*Am*
Cell Parameters				
a	13.089	9.879	9.865	7.549
b	13.047	14.139	14.300	14.740
c	13.218	8.693	8.668	13.072
α				90
β		124.81	124.2	90
γ				111.9
Vol.	2257.3	996.94	1011.35	1349.6
Ref.	Alberti et al.(1981)	Rinaldi et al. (1974)	Rinaldi et al.(1974)	Schramm & Fischer (1971)

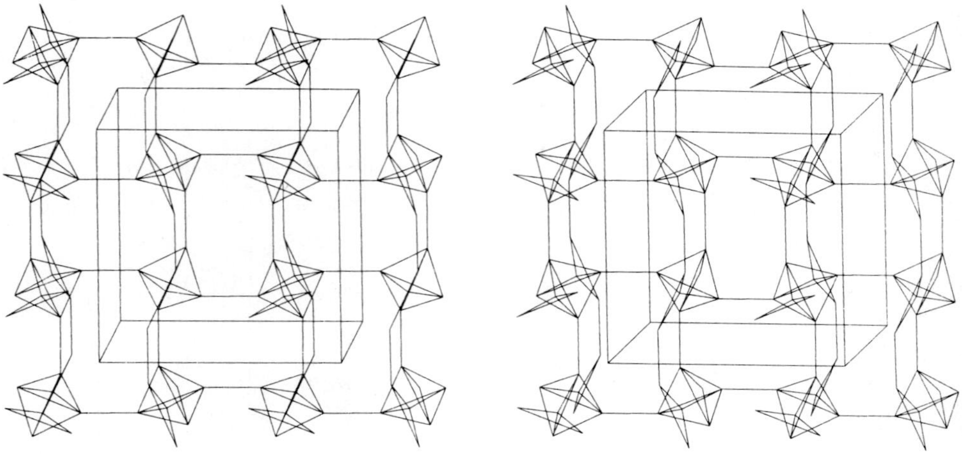

Figure 8.7e. Thomsonite, $NaCa_2Al_5Si_5O_{20}6H_2O$, stereoscopic c-axis projection. The framework is shown schematically.

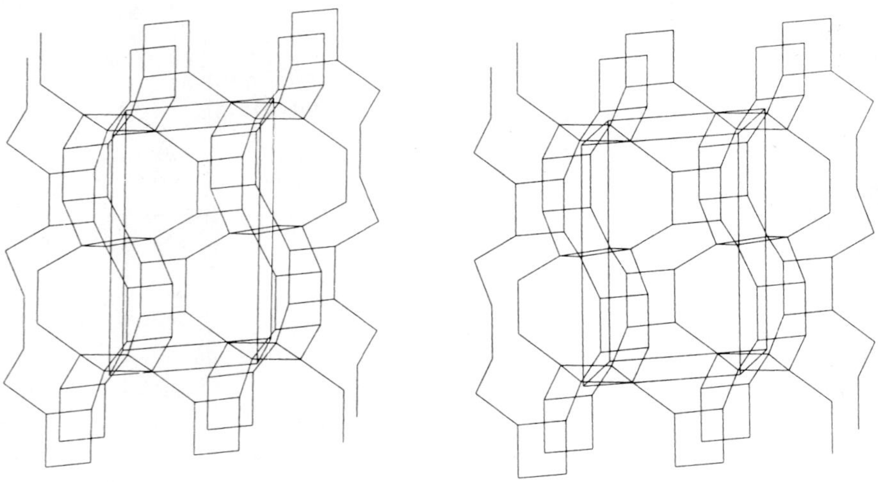

Figure 8.7f. Phillipsite, $K_2(Ca,Na)_2Al_5Si_{10}O_{32}12H_2O$, stereoscopic c-axis projection. The framework is shown schematically. Harmotome has a very similar framework.

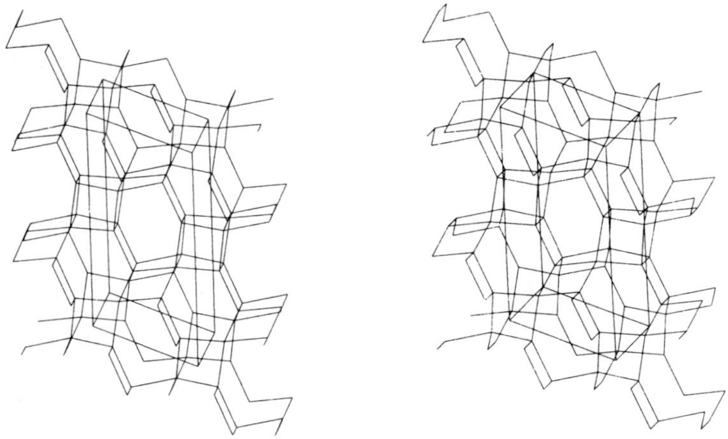

Figure 8.7g. Laumontite, $CaAl_2Si_4O_{12} \cdot 4H_2O$, stereoscopic c-axis projection. The framework is shown schematically.

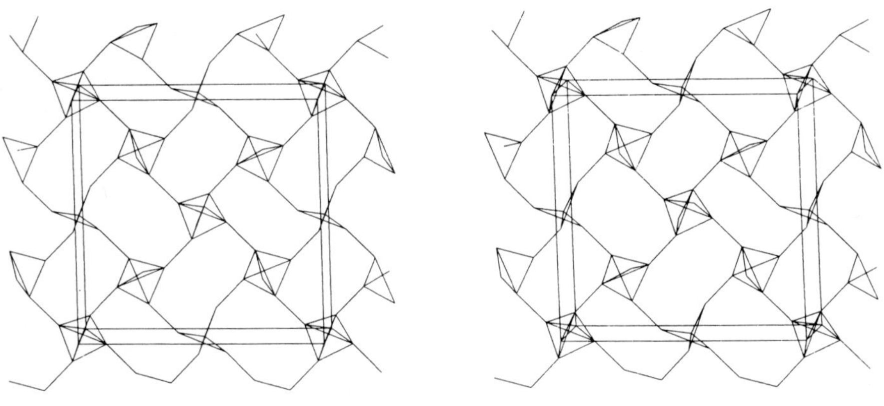

Figure 8.7h. Natrolite, $Na_2Al_2Si_3O_{10} \cdot 2H_2O$, stereoscopic c-axis projection. The framework is shown schematically.

Table 8.7.1d. Zeolite Group Unit Cells (continued).

End-Member	Natrolite	Sodalite	Stilbite
Formula	Na_2 $Al_2Si_3O_{10}$ $2H_2O$	Na_4 $Al_3Si_3O_{12}$ Cl	$Na_{1.28}Ca_{4.18}Mg_{.18}$ $Al_{10.30}Si_{25.71}O_{72}$ $34H_2O$
Form.Wt.	380.23	484.61	2968.
Density	2.238	2.306	2.23
Mol Volume	169.87	210.16	1331.
Z	8	2	1
Cryst.Sys.	Orthorhombic	Isometric	Monoclinic
Cryst.Class	$mm2$	$\bar{4}3n$	$2/m$
Space Group	$Fdd2$	$P\bar{4}3n$	$C2/m$
Cell Parameters			
a	18.326	8.870	13.64
b	18.652	8.870	18.24
c	6.601	8.870	11.27
β			128.0
Vol.	2256.3	697.86	2210.
Ref.	Pechar et al. (1983)	Löns & Schulz(1967)	Galli (1971)

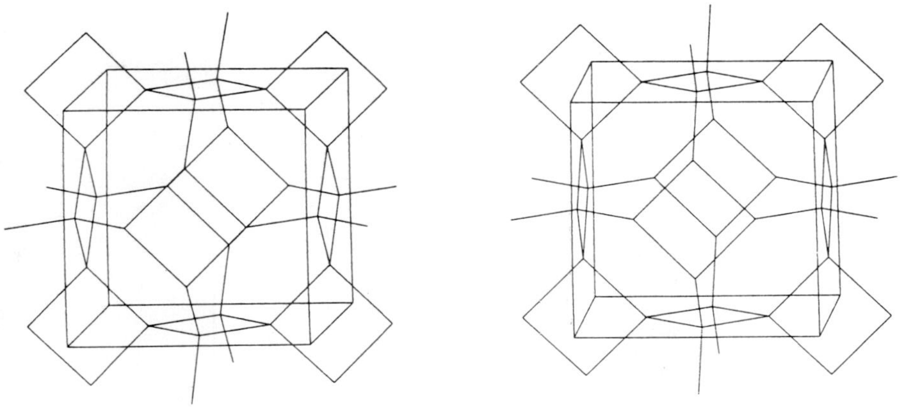

Figure 8.7i. Sodalite, $Na_4Al_3Si_3O_{12}Cl$, stereoscopic c-axis projection. The framework is shown schematically.

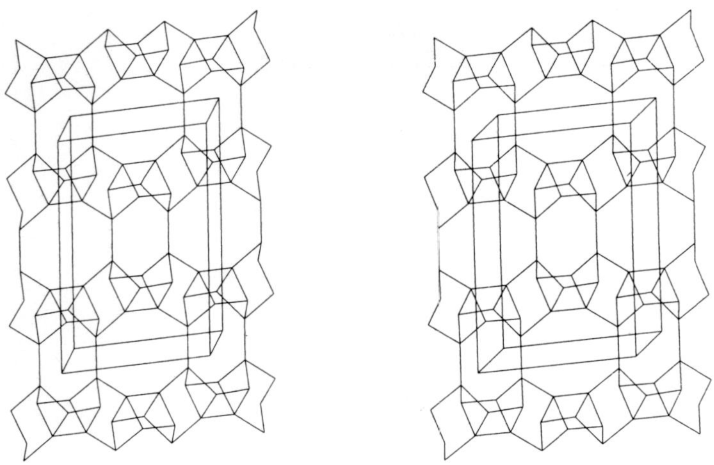

Figure 8.7j. Stilbite, $NaCa_2Al_5Si_{13}O_{36}16H_2O$, stereoscopic c-axis projection. The framework is shown schematically.

Table 8.7.2a. Zeolite Group Tetrahedral Sites.

End-Member	Analcime		Chabazite
Site	T1	T2	T
C.N.	4	4	4
Occupant	$Al_{.27}Si_{.73}$	$Al_{.45}Si_{.55}$	$Al_{.31}Si_{.69}$
Point Sym.	1	2	1
Wyckoff Not.	32g	16f	12i
Frac.Coord.			
x	.1244	.1623	.1044
y	.1624	.4123	.3338
z	.4124	1/8	.8749
Distances			
Mean	1.639	1.663	1.647
σ	.006	.008	.005
Poly.Vol.	2.252	2.346	2.287
Q.E.	1.0026	1.0036	1.0016
Ang.Var.	10.6	14.6	6.46
Site Energy	-3862.	-3517.	-3750.
Model Charge	3.73	3.55	3.69

Table 8.7.2b. Zeolite Group Tetrahedral Sites (continued).

End-Member	K-Mordenite			
Site	T1	T2	T3	T4
C.N.	4	4	4	4
Occupant	$Al_{.10}Si_{.90}$	$Al_{.16}Si_{.84}$	$Al_{.25}Si_{.75}$	$Al_{.18}Si_{.82}$
Point Sym.	1	1	*m*	*m*
Wyckoff Not.	16h	16h	8g	8g
Frac.Coord.				
x	.19850	.19656	.08755	.08671
y	.42742	.19071	.38181	.22693
z	.54128	.54487	1/4	1/4
Distances				
Mean	1.619	1.611	1.634	1.629
σ	.009	.012	.027	.062
Poly.Vol.	2.174	2.142	2.237	2.212
Q.E.	1.0013	1.0011	1.0013	1.0020
Ang.Var.	5.27	4.81	5.03	8.17
Site Energy	-4231.	-4087.	-3871.	-4019.
Model Charge	3.90	3.84	3.75	3.82

Table 8.7.2c. Zeolite Group Tetrahedral Sites (continued).

End-Member	Clinoptilolite				
Site	T1	T2	T3	T4	T5
C.N.	4	4	4	4	4
Occupant	$Al_{.17}Si_{.83}$	$Al_{.31}Si_{.69}$	$Al_{.13}Si_{.87}$	$Al_{.11}Si_{.89}$	$Al_{.10}Si_{.90}$
Point Sym.	1	1	1	1	2
Wyckoff Not.	8j	8j	8j	8j	4g
Frac.Coord.					
x	.17906	.21334	.20846	.06623	0
y	.16943	.41099	.19034	.29837	.21651
z	.0963	.5040	.7153	.4148	0
Distances					
Mean	1.624	1.645	1.618	1.615	1.614
σ	.005	.011	.011	.004	.003
Poly.Vol.	2.195	2.281	2.169	2.159	2.154
Q.E.	1.0010	1.0012	1.0010	1.0010	1.0013
Ang.Var.	4.49	4.47	3.68	4.15	5.34
Site Energy	-4030.	-3790.	-4075.	-4199.	-4272.
Model Charge	3.83	3.69	3.87	3.89	3.90

Table 8.7.2d. Zeolite Group Tetrahedral Sites (continued).

End-Member	Heulandite				
Site	T1	T2	T3	T4	T5
C.N.	4	4	4	4	4
Occupant	$Al_{.27}Si_{.73}$	$Al_{.40}Si_{.60}$	$Al_{.17}Si_{.83}$	$Al_{.19}Si_{.81}$	$Al_{.27}Si_{.73}$
Point Sym.	1	1	1	1	2
Wyckoff Not.	8j	8j	8j	8j	4g
Frac.Coord.					
x	.1798	.2119	.2078	.0643	0
y	.1689	.4097	.1911	.2992	.2127
z	.0968	.4996	.7169	.4107	0
Distances					
Mean	1.635	1.657	1.623	1.629	1.636
σ	.009	.012	.006	.014	.015
Poly.Vol.	2.241	2.327	2.190	2.214	2.243
Q.E.	1.0007	1.0028	1.0015	1.0010	1.0016
Ang.Var.	3.02	10.58	5.93	4.35	6.21
Site Energy	-3878.	-3594.	-4010.	-4034.	-3961.
Model Charge	3.73	3.60	3.83	3.81	3.73

Table 8.7.2e. Zeolite Group Tetrahedral Sites (continued).

End-Member	K-Heulandite				
Site	T1	T2	T3	T4	T5
C.N.	4	4	4	4	4
Occupant	$Al_{.22}Si_{.78}$	$Al_{.40}Si_{.60}$	$Al_{.19}Si_{.81}$	$Al_{.13}Si_{.87}$	$Al_{.30}Si_{.70}$
Point Sym.	1	1	1	1	2
Wyckoff Not.	8j	8j	8j	8j	4g
Frac.Coord.					
x	.1794	.2146	.2083	.0668	0
y	.1686	.4108	.1912	.2983	.2173
z	.0978	.5063	.7161	.4176	0
Distances					
Mean	1.631	1.657	1.628	1.619	1.640
σ	.012	.013	.005	.010	.013
Poly.Vol.	2.220	2.332	2.206	2.174	2.259
Q.E.	1.0016	1.0012	1.0022	1.0018	1.0018
Ang.Var.	7.10	3.89	8.65	7.39	7.31
Site Energy	-4005.	-3656.	-4048.	-4130.	-3844.
Model Charge	3.78	3.60	3.81	3.87	3.70

Table 8.7.2f. Zeolite Group Tetrahedral Sites (continued).

End-Member	Thomsonite					
Site	Si1	Al1	Si2	Al2	Si3	Al3
C.N.	4	4	4	4	4	4
Occupant	Si	Al	Si	Al	Si	Al
Point Sym.	2	2	1	1	1	1
Wyckoff Not.	4d	4d	8e	8e	8e	8e
Frac.Coord.						
x	1/4	1/4	.1130	.1196	.3081	.3095
y	1/4	3/4	.6958	.3054	.3848	.6238
z	.6883	.6908	.5006	.4965	.3781	.3806
Distances						
Mean	1.617	1.741	1.626	1.743	1.619	1.741
σ	.001	.002	.007	.011	.006	.007
Poly.Vol.	2.169	2.703	2.200	2.705	2.169	2.688
Q.E.	1.0006	1.0012	1.0015	1.0027	1.0027	1.0050
Ang.Var.	2.57	5.03	6.25	11.04	10.50	19.38
Site Energy	-4297.	-2422.	-4190.	-2447.	-4333.	-2464.
Model Charge	4.0	3.0	4.0	3.0	4.0	3.0

Table 8.7.2g. Zeolite Group Tetrahedral Sites (continued).

End-Member	Harmotome			
Site	T1	T2	T3	T4
C.N.	4	4	4	4
Occupant	$Al_{.31}Si_{.69}$	$Al_{.31}Si_{.69}$	$Al_{.31}Si_{.69}$	$Al_{.31}Si_{.69}$
Point Sym.	1	1	1	1
Wyckoff Not.	4f	4f	4f	4f
Frac.Coord.				
x	.7367	4214	.0577	.1216
y	.0248	.1410	.0075	.1390
z	.2840	.0136	.2898	.0375
Distances				
Mean	1.653	1.640	1.640	1.645
σ	.011	.007	.009	.005
Poly.Vol.	2.308	2.258	2.260	2.279
Q.E.	1.0029	1.0011	1.0017	1.0018
Ang.Var.	12.00	4.74	6.68	6.90
Site Energy	-3777.	-3846.	-3820.	-3818.
Model Charge	3.69	3.69	3.69	3.69

Table 8.7.2h. Zeolite Group Tetrahedral Sites (continued).

End-Member	Phillipsite			
Site	T1	T2	T3	T4
C.N.	4	4	4	4
Occupant	$Al_{.33}Si_{.67}$	$Al_{.33}Si_{.67}$	$Al_{.33}Si_{.67}$	$Al_{.33}Si_{.67}$
Point Sym.	1	1	1	1
Wyckoff Not.	4f	4f	4f	4f
Frac.Coord.				
x	.7362	.4206	.0604	.1204
y	.0248	.1409	.0078	.1396
z	.2805	.0019	.2844	.0421
Distances				
Mean	1.651	1.654	1.664	1.654
σ	.008	.008	.011	.010
Poly.Vol.	2.300	2.319	2.354	2.315
Q.E.	1.0029	1.0016	1.0028	1.0015
Ang.Var.	11.94	6.62	11.40	6.09
Site Energy	-3708.	-3718.	-3701.	-3759.
Model Charge	3.67	3.67	3.67	3.67

Table 8.7.2i. Zeolite Group Tetrahedral Sites (continued).

End-Member	Laumontite					
Site	T1	T2	T3	T4	T5	T6
C.N.	4	4	4	4	4	4
Occupant	Si	$Al_{.68}Si_{.32}$	$Si_{.92}Al_{.08}$	$Si_{.85}Al_{.15}$	$Al_{.94}Si_{.06}$	$Si_{.85}Al_{.15}$
Point Sym.	1	1	1	1	1	1
Wyckoff Not.	4b	4b	4b	4b	4b	4b
Frac.Coord.						
x	.3310	.7565	.1638	.8152	.2256	.6465
y	.2594	.3700	.4169	.5779	.6282	.7361
z	.3814	.3082	.3833	.3828	.3098	.3823
Distances						
Mean	1.603	1.712	1.614	1.631	1.740	1.631
σ	.016	.072	.034	.015	.023	.015
Poly.Vol.	2.109	2.533	2.139	2.220	2.672	2.222
Q.E.	1.0016	1.0122	1.0061	1.0027	1.0083	1.0012
Ang.Var.	6.19	44.24	25.07	10.60	30.87	4.95
Site Energy	-4141.	-2718.	-3754.	-3553.	-2288.	-3647.
Model Charge	4.00	3.32	3.92	3.85	3.06	3.85

Table 8.7.2j. Zeolite Group Tetrahedral Sites (continued).

End-Member	Natrolite			Sodalite	
Site	T1	T2	T3	Si	Al
C.N.	4	4	4	4	4
Occupant	Si	Si	Al	Si	Al
Point Sym.	2	1	1	$\bar{4}$	$\bar{4}$
Wyckoff Not.	8a	16b	16b	6d	6c
Frac.Coord.					
x	0	.15330	.03780	1/4	1/4
y	0	.21110	.09380	0	1/2
z	0	.62370	.61580	1/2	0
Distances					
Mean	1.621	1.625	1.737	1.628	1.728
σ	.007	.018	.004	0	0
Poly.Vol.	2.184	2.201	2.682	2.207	2.648
Q.E.	1.0010	1.0008	1.0022	1.0019	1.0005
Ang.Var.	4.25	3.29	8.88	7.56	2.30
Site Energy	-4228.	-4291.	-2476.	-4361.	-2526.
Model Charge	4.0	4.0	3.0	4.0	3.0

Table 8.7.2k. Zeolite Group Tetrahedral Sites (continued).

End-Member	Stilbite				
Site	T1	T2	T3	T4	T5
C.N.	4	4	4	4	4
Occupant	$Si_{.71}Al_{.29}$	$Si_{.71}Al_{.29}$	$Si_{.71}Al_{.29}$	$Si_{.71}Al_{.29}$	$Si_{.71}Al_{.29}$
Point Sym.	1	1	1	1	2
Wyckoff Not.	8j	8j	8j	8j	4g
Frac.Coord.					
x	.4830	.2653	.1892	.1124	0
y	.3042	.3097	.0893	.3166	.2610
z	.2420	.2619	.4846	.5013	0
Distances					
Mean	1.647	1.637	1.645	1.638	1.630
σ	.021	.009	.014	.011	.019
Poly.Vol.	2.291	2.247	2.282	2.253	2.216
Q.E.	1.0011	1.0015	1.0011	1.0012	1.0020
Ang.Var.	4.53	6.51	4.01	5.10	7.94
Site Energy	-3868.	-3871.	-3941.	-3821.	-3980.
Model Charge	3.71	3.71	3.71	3.71	3.71

Chapter 9

CARBONATES, NITRATES, SULFATES AND PHOSPHATES

Table 9.1.1. Calcite Group Unit Cells.

End-Member	Magnesite	Smithsonite	Siderite	Rhodochrosite	Otavite	Calcite	Soda Niter
Formula	$MgCO_3$	$ZnCO_3$	$FeCO_3$	$MnCO_3$	$CdCO_3$	$CaCO_3$	$NaNO_3$
Form.Wt.	84.321	125.379	115.856	114.947	172.409	100.089	84.995
Density	3.010	4.434	3.937	3.720	5.024	2.7106	2.261
Mol Volume	28.012	28.276	29.429	30.904	34.316	36.9257	37.594
Z	6	6	6	6	6	6	6
Cryst.Sys.	Hexagonal-R	Hexagonal-R	Hexagonal-R	Hexagonal-R	Hexagonal-R	Hexagonal-R	Hexagonal-R
Cryst.Class	$\bar{3}m$	$\bar{3}m$	$\bar{3}m$	$\bar{3}m$	$\bar{3}m$	$\bar{3}m$	$\bar{3}m$
Space Group	$R\bar{3}c$	$R\bar{3}c$	$R\bar{3}c$	$R\bar{3}c$	$R\bar{3}c$	$R\bar{3}c$	$R\bar{3}c$
Cell Parameters							
a	4.6328	4.6526	4.6916	4.7682	4.923	4.9896	5.0708
c	15.0129	15.0257	15.3796	15.6354	16.287	17.0610	16.818
Vol.	279.05	281.68	293.17	307.86	341.85	367.85	374.51
Ref.	Effenberger et al.(1981)	Effenberger et al.(1981)	Effenberger et al.(1981)	Effenberger et al.(1981)	Borodin et al.(1979)	Effenberger et al.(1981)	Sass et al. (1957)

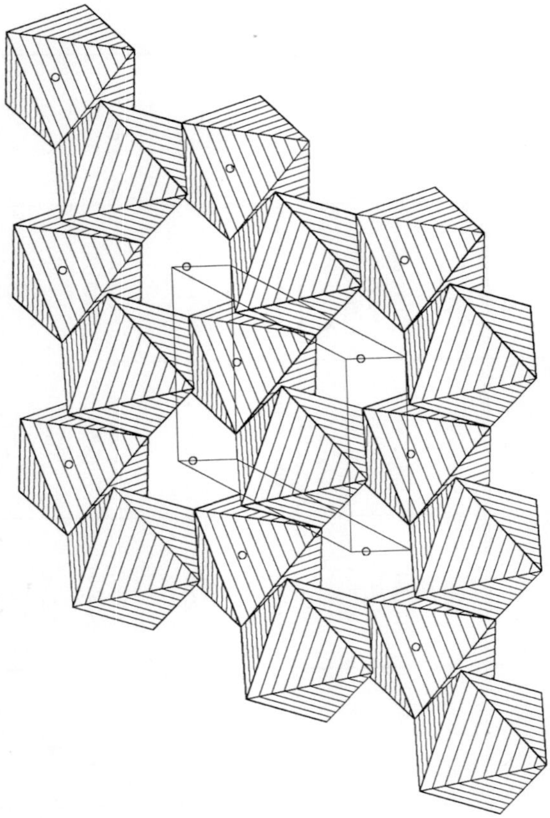

Figure 9.1. Calcite, $CaCO_3$, perspective c-axis projection, b vertical. There is a single Ca site in octahedral coordination with point symmetry 3, and a single C site (circle) in triangular three-coordination and point symmetry 32. The rhombohedral carbonates and nitrates; magnesite ($MgCO_3$), smithsonite ($ZnCO_3$), siderite ($FeCO_3$), rhodochrosite ($MnCO_3$), otavite ($CdCO_3$), and soda niter ($NaNO_3$); are all isostructural. Two layers of octahedra are shown. In the ordered variant dolomite, the Ca and Mg occupy alternate layers of octahedra to reduce the symmetry to $R\bar{3}$ from $R\bar{3}c$ of calcite.

Table 9.1.2. Calcite Group Octahedral Sites.

End-Member	Magnesite	Smithsonite	Siderite	Rhodochrosite	Otavite	Calcite	Soda Niter
C.N.	6	6	6	6	6	6	6
Occupant	Mg	Zn	Fe^{+2}	Mn^{+2}	Cd	Ca	Na
Point Sym.	$\bar{3}$	$\bar{3}$	$\bar{3}$	$\bar{3}$	$\bar{3}$	$\bar{3}$	$\bar{3}$
Wyckoff Not.	6b	6b	6b	6b	6b	6b	6b
Frac.Coord.							
x	0	0	0	0	0	0	0
y	0	0	0	0	0	0	0
z	0	0	0	0	0	0	0
Distances							
O1 (6)	2.102	2.111	2.144	2.190	2.288	2.360	2.417
Poly.Vol.	12.360	12.520	13.122	13.986	15.944	17.468	18.828
Q.E.	1.0010	1.0008	1.0013	1.0009	1.0008	1.0019	1.0000
Ang.Var.	3.55	2.80	4.59	3.42	2.80	7.09	0.06
Site Energy	-1067.	-1061.	-1039.	-1008.	-1120.	-1066.	-295.

Table 9.1.3. Calcite Group Triangular Sites.

End-Member	Magnesite	Smithsonite	Siderite	Rhodochrosite	Otavite	Calcite	Soda Niter
C.N.	3	3	3	3	3	3	3
Occupant	C	C	C	C	C	C	N
Point Sym.	32	32	32	32	32	32	32
Wyckoff Not.	6a	6a	6a	6a	6a	6a	6a
Frac.Coord.							
x	0	0	0	0	0	0	0
y	0	0	0	0	0	0	0
z	1/4	1/4	1/4	1/4	1/4	1/4	1/4
Distances							
O1 (3)	1.285	1.286	1.287	1.287	1.290	1.281	1.218
Site Energy	-5179.	-5178.	-5170.	-5171.	-5159.	-5193.	-7741.

Table 9.2.1. Dolomite Group Unit Cells.

End-Member	Dolomite	Ankerite
Formula	$CaMg(CO_3)_2$	$CaFe(CO_3)_2$
Form.Wt.	184.411	215.946
Density	2.868	3.293
Mol Volume	64.293	65.576
Z	3	3
Cryst.Sys.	Hexagonal-R	Hexagonal-R
Cryst.Class	$\bar{3}$	$\bar{3}$
Space Group	$R\bar{3}$	$R\bar{3}$
Cell Parameters		
a	4.8069	4.830
c	16.0034	16.167
Vol.	320.24	326.63
Ref.	Reeder (1983)	Beran & Zemann (1977)

Table 9.2.2. Dolomite Group M Sites.

End-Member	Dolomite		Ankerite	
Site	A	B	A	B
C.N.	6	6	6	6
Occupant	Ca	Mg	Ca	Mg
Point Sym.	$\bar{3}$	$\bar{3}$	$\bar{3}$	$\bar{3}$
Wyckoff Not.	3a	3b	3a	3b
Frac.Coord.				
x	0	0	0	0
y	0	0	0	0
z	0	1/2	0	1/2
Distances				
O1 (6)	2.381	2.084	2.371	2.126
Poly.Vol.	17.950	12.060	17.726	12.789
Q.E.	1.0016	1.0008	1.0020	1.0009
Ang.Var.	5.84	3.08	7.33	3.44
Site Energy	-936.	-1175.	-944.	-1141.

Table 9.2.3. Dolomite Group Triangular Sites.

End-Member	Dolomite	Ankerite
C.N.	3	3
Occupant	C	C
Point Sym.	3	3
Wyckoff Not.	6c	6c
Frac.Coord.		
x	0	0
y	0	0
z	.24282	.2442
Distances		
O1 (3)	1.284	1.284
Site Energy	-5178.	-5178.

Table 9.3.1. Aragonite Group Unit Cells.

End-Member	Aragonite	Strontianite	Cerussite	Witherite	Niter
Formula	$CaCO_3$	$SrCO_3$	$PbCO_3$	$BaCO_3$	KNO_3
Form.Wt.	100.089	147.629	267.199	197.349	101.107
Density	2.930	3.843	6.577	4.314	2.079
Mol Volume	34.166	38.416	40.629	45.745	48.643
Z	4	4	4	4	4
Cryst.Sys.	Orthorhombic	Orthorhombic	Orthorhombic	Orthorhombic	Orthorhombic
Cryst.Class	*mmm*	*mmm*	*mmm*	*mmm*	*mmm*
Space Group	*Pmcn*	*Pmcn*	*Pmcn*	*Pmcn*	*Pmcn*
Cell Parameters					
a	4.9614	5.090	5.180	5.3126	5.4119
b	7.9671	8.358	8.492	8.8958	9.1567
c	5.7404	5.997	6.134	6.4284	6.5189
Vol.	226.91	255.13	269.83	303.81	323.05
Ref.	De Villiers (1971)	De Villiers (1971)	Sahl (1974)	De Villiers (1971)	Nimmo & Lucas (1973)

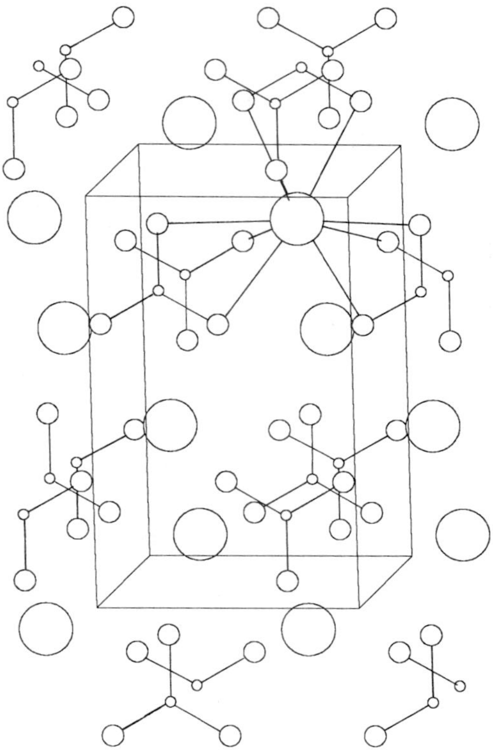

Figure 9.3. Aragonite, $CaCO_3$, perspective c-axis projection, b vertical. Ca is in irregular nine-coordination with point symmetry m, and C is in triangular three-coordination also with point symmetry m. The orthorhombic carbonates and nitrates; strontianite ($SrCO_3$), cerussite ($PbCO_3$), witherite ($BaCO_3$), and niter (KNO_3); are isostructural.

Table 9.3.2. Aragonite Group M Sites.

End-Member	Aragonite	Strontianite	Cerussite	Witherite	Niter
C.N.	9	9	9	9	9
Occupant	Ca	Sr	Pb	Ba	K
Point Sym.	*m*	*m*	*m*	*m*	*m*
Wyckoff Not.	4c	4c	4c	4c	4c
Frac.Coord.					
x	1/4	1/4	1/4	1/4	1/4
y	.4150	.4160	.4171	.41631	.4166
z	.7597	.7569	.7551	.7549	.7568
Distances					
O1(1)	2.419	2.552	2.621	2.737	2.867
O1(2)	2.653	2.725	2.769	2.868	2.928
O2(2)	2.550	2.666	2.712	2.836	2.888
O2(2)	2.445	2.561	2.677	2.742	2.849
O2(2)	2.520	2.634	2.656	2.816	2.888
Mean	2.528	2.636	2.695	2.807	2.886
σ	.085	.067	.051	.053	.029
Poly.Vol.	37.54	42.48	45.47	50.90	54.80
Site Energy	-954.	-906.	-878.	-841.	-242.

Table 9.3.3. Aragonite Group Triangular Sites.

End-Member	Aragonite	Strontianite	Cerussite	Witherite	Niter
C.N.	3	3	3	3	3
Occupant	C	C	C	C	N
Point Sym.	*m*	*m*	*m*	*m*	*m*
Wyckoff Not.	4c	4c	4c	4c	4c
Frac.Coord.					
x	1/4	1/4	1/4	1/4	1/4
y	.7622	.7601	.754	.7570	.7548
z	-.0862	-.0864	-.092	-.0810	-.0848
Distances					
O1(1)	1.278	1.270	1.30	1.283	1.240
O2(2)	1.284	1.293	1.25	1.289	1.246
Mean	1.282	1.285	1.27	1.287	1.244
σ	.003	.014	.03	.004	.003
Site Energy	-5015.	-5032.	-5120.	-5074.	-7368.

Table 9.4.1. Barite Group Unit Cells.

End-Member	Celestite	Anglesite	Barite
Formula	$SrSO_4$	$PbSO_4$	$BaSO_4$
Form.Wt.	183.682	303.252	233.402
Density	3.961	6.321	4.467
Mol Volume	46.371	47.977	52.245
Z	4	4	4
Cryst.Sys.	Orthorhombic	Orthorhombic	Orthorhombic
Cryst.Class	*mmm*	*mmm*	*mmm*
Space Group	*Pbnm*	*Pbnm*	*Pbnm*
Cell Parameters			
a	6.870	6.959	7.157
b	8.371	8.482	8.884
c	5.355	5.398	5.457
Vol.	307.96	318.62	346.97
Ref.	Miyake et al.(1978)	Miyake et al.(1978)	Miyake et al.(1978)

Figure 9.4. Barite, $BaSO_4$, perspective c-axis projection, b vertical. Ba is in irregular ten-coordination with point symmetry m, and S is in a highly regular tetrahedron, also with point symmetry m. Celestite ($SrSO_4$) and anglesite ($PbSO_4$) also have this structure.

Table 9.4.2. Barite Group M Sites.

End-Member	Celestite	Anglesite	Barite
C.N.	10	10	10
Occupant	Sr	Pb	Ba
Point Sym.	*m*	*m*	*m*
Wyckoff Not.	4c	4c	4c
Frac.Coord.			
x	.1583	.1667	.1585
y	.1839	.1879	.1845
z	1/4	1/4	1/4
Distances Cation-Ox			
O1(1)	2.536	2.613	2.770
O2(2)	2.980	3.005	3.075
O2(1)	2.636	2.619	2.798
O3(2)	2.690	2.648	2.811
O3(2)	2.653	2.737	2.811
O3(2)	2.816	2.910	2.911
Mean	2.745	2.783	2.879
σ	.149	.159	.113
Poly.Vol.	45.90	47.97	52.99
Site Energy	-953.	-939.	-892.

Table 9.4.3. Barite Group S Sites.

End-Member	Celestite	Anglesite	Barite
C.N.	4	4	4
Occupant	S	S	S
Point Sym.	*m*	*m*	*m*
Wyckoff Not.	4c	4c	4c
Frac.Coord.			
x	.1853	.1842	.1911
y	.4382	.4367	.4375
z	1/4	1/4	1/4
Distances			
O1(1)	1.452	1.453	1.468
O2(1)	1.472	1.481	1.470
O3(2)	1.486	1.497	1.486
Mean	1.474	1.482	1.478
σ	.016	.021	.010
Poly.Vol.	1.643	1.669	1.655
Q.E.	1.0005	1.0012	1.0004
Ang.Var.	2.31	5.20	2.01
Site Energy	-9648.	-9590.	-9622.

Table 9.5.1. Gypsum and Anhydrite Unit Cells.

End-Member	Gypsum	Anhydrite
Formula	$CaSO_4{\cdot}2H_2O$	$CaSO_4$
Form.Wt.	172.173	136.142
Density	2.313	2.953
Mol Volume	74.440	46.103
Z	4	4
Cryst.Sys.	Monoclinic	Orthorhombic
Cryst.Class	$2/m$	mmm
Space Group	$I2/a$	$Amma$
Cell Parameters		
a	5.670	7.006
b	15.201	6.998
c	6.533	6.245
β	118.60	90
Vol.	494.37	306.18
Ref.	Cole & Lancucki(1974)	Kirfel & Will (1980)

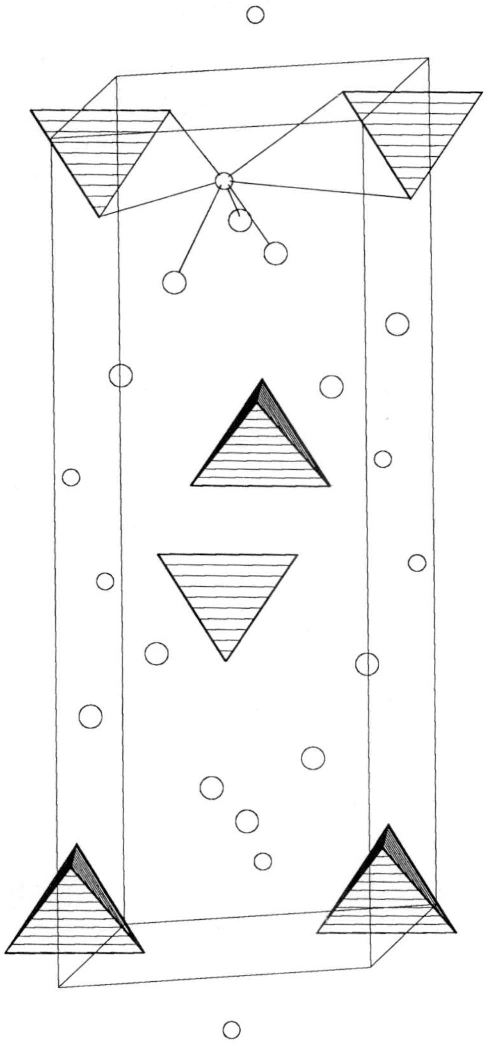

Figure 9.5a. Gypsum, $CaSO_4 2H_2O$, perspective c-axis projection, b vertical. Ca is in eight-coordination with point symmetry m, and S is in a highly regular tetrahedron with point symmetry 2.

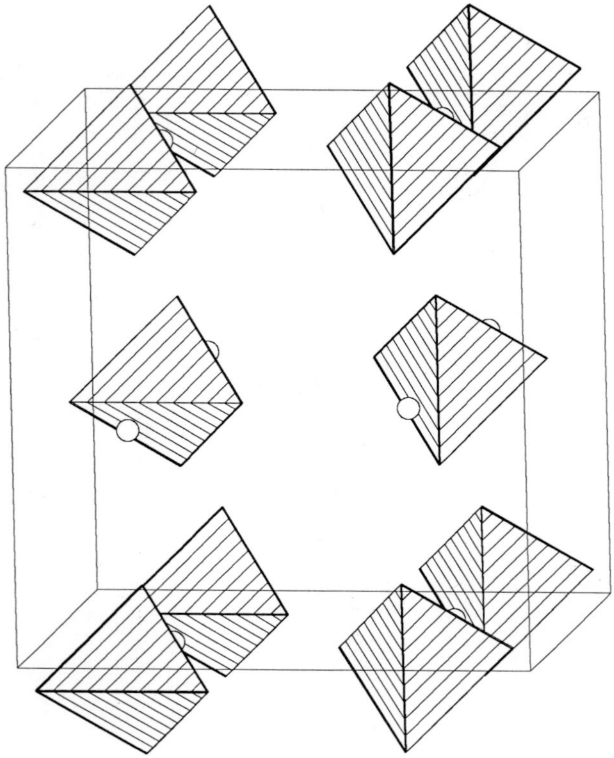

Figure 9.5b. Anhydrite, $CaSO_4$, perspective c-axis projection, b vertical. Ca is in eight-coordination with point symmetry $m2m$, and S is in a highly regular tetrahedron also with point symmetry $m2m$.

Table 9.5.2. Gypsum and Anhydrite Ca Sites.

End-Member	Gypsum	Anhydrite
C.N.	8	8
Occupant	Ca	Ca
Point Sym.	2	$m2m$
Wyckoff Not.	4e	4c
Frac.Coord.		
x	1/2	3/4
y	.0797	0
z	1/4	.34765
Distances		
O1(2)	2.528	2.563
O1(2)		2.464
O2(2)	2.545	2.347
O2(2)	2.378	2.510
O(w)(2)	2.380	
Mean	2.458	2.471
σ	.084	.085
Poly.Vol.	25.983	26.433
Site Energy	-978.	-1014.

Table 9.5.3. Gypsum and Anhydrite S Sites.

End-Member	Gypsum	Anhydrite
C.N.	4	4
Occupant	S	S
Point Sym.	2	*m* 2*m*
Wyckoff Not.	4e	4c
Frac.Coord.		
x	0	1/4
y	0.0772	0
z	3/4	.15577
Distances		
O1(2)	1.457	1.473
O2(2)	1.461	1.472
Mean	1.459	1.472
σ	.002	.000
Poly.Vol.	1.589	1.635
Q.E.	1.0023	1.0012
Ang.Var.	9.7	4.8
Site Energy	-9404.	-9651.

Table 9.6.1. Apatite Group Unit Cells.

End-Member	Hydroxyapatite	Fluorapatite	Chlorapatite
Formula	$Ca_5(PO_4)_3OH$	$Ca_5(PO_4)_3F$	$Ca_5(PO_4)_3Cl$
Form.Wt.	502.322	504.313	520.767
Density	3.153	3.201	3.185
Mol Volume	159.334	157.527	163.527
Z	2	2	2
Cryst.Sys.	Hexagonal	Hexagonal	Hexagonal
Cryst.Class	$6/m$	$6/m$	$6/m$
Space Group	$P6_3/m$	$P6_3/m$	$P6_3/m$
Cell Parameters			
a	9.424	9.367	9.628
c	6.879	6.884	6.764
Vol.	529.09	523.09	543.01
Ref.	Sudarsanan and Young (1969)	Sudarsanan et al.(1972)	Mackie et al. (1972)

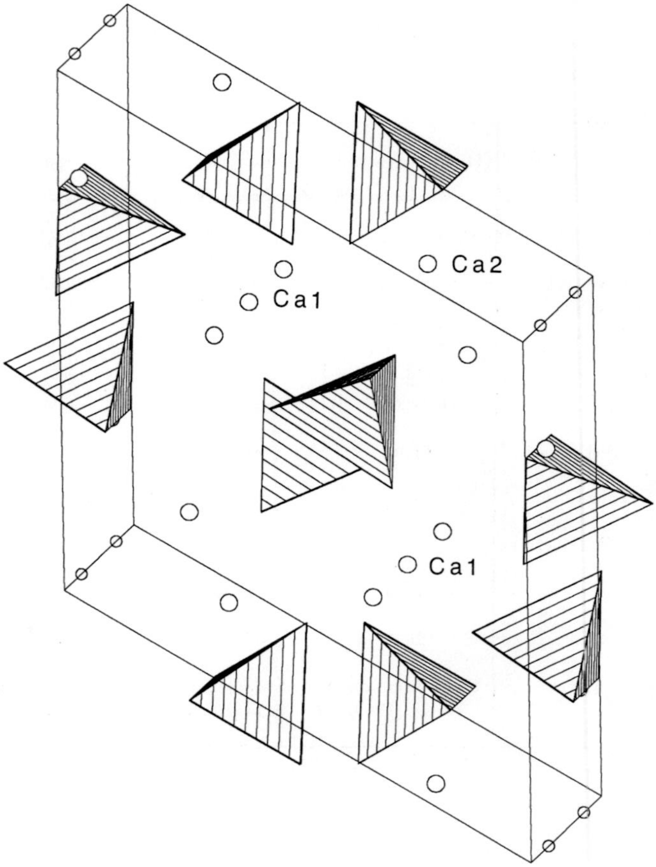

Figure 9.6. Apatite, $Ca_5(PO_4)_3OH$, perspective c-axis projection, b vertical. There are two distinct Ca sites; Ca1 is in nine-coordination with point symmetry 3 and Ca2 is in 8 coordination with point symmetry m. There is a single P site in tetrahedral coordination and point symmetry m.

Table 9.6.2. Apatite Group Ca Sites.

End-Member	Hydroxyapatite		Fluorapatite		Chlorapatite	
Site	Ca1	Ca2	Ca1	Ca2	Ca1	Ca2
C.N.	9	8	9	7	9	8
Point Sym.	3	*m*	3	*m*	3	*m*
Wyckoff Not.	4f	6h	4f	6h	4f	6h
Frac.Coord.						
x	1/3	.2465	1/3	.2416	1/3	.2596
y	2/3	.9931	2/3	.0071	2/3	.0053
z	.0013	1/4	.0012	1/4	.0038	1/4
Distances						
O1	(3)2.407	(1)2.705	(3)2.397	(1)2.814	(3)2.398	(1)2.973
O2	(3)2.452	(1)2.357	(3)2.453	(1)2.384	(3)2.448	(1)2.295
O3	(3)2.805	(2)2.344	(3)2.801	(2)2.384	(3)2.789	(2)2.534
O3		(2)2.512		(2)2.398		(2)2.336
OH/F/Cl		(2)2.383		(1)2.231		(2)2.801
Mean	2.555	2.443	2.550	2.428	2.545	2.577
σ	.189	.127	.190	.180	.184	.255
Poly.Vol.	31.974	26.138	31.750	21.506	31.977	34.126
Site Energy	-969.	-1009.	-984.	-996.	-984.	-976.

Table 9.6.3. Apatite Group P Sites.

End-Member	Hydroxyapatite	Fluorapatite	Chlorapatite
C.N.	4	4	4
Point Sym.	*m*	*m*	*m*
Wyckoff Not.	6h	6h	6h
Frac.Coord.			
x	.3983	.3981	.4078
y	.3683	.3688	.3754
z	1/4	1/4	1/4
Distances			
O1	1.537	1.534	1.536
O2	1.545	1.541	1.542
O3(2)	1.529	1.534	1.529
Mean	1.535	1.536	1.534
σ	.008	.003	.006
Poly.Vol.	1.853	1.857	1.849
Q.E.	1.0011	1.0008	1.0017
Ang.Var.	4.3	3.7	7.1
Site Energy	-6831.	-6835.	-6834.

Table 9.7.1. Monazite Unit Cell

End-Member	Monazite
Formula	$CePO_4$
Form.Wt.	235.09
Density	5.226
Mol Volume	44.98
Z	4
Cryst.Sys.	Monoclinic
Cryst.Class	$2/m$
Space Group	$P\,2_1/n$
Cell Parameters	
a	6.77
b	7.04
c	6.46
β	104.0
Vol.	298.7
Ref.	Ghouse(1968)

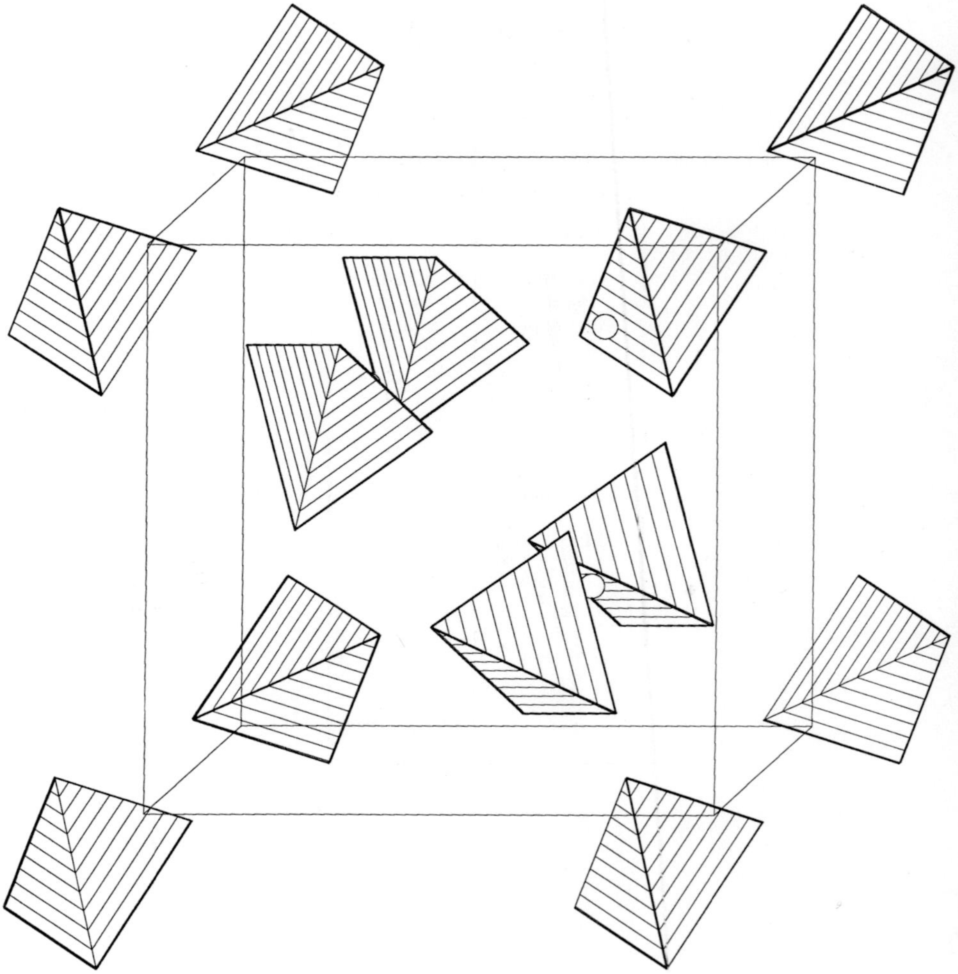

Figure 9.7. Monazite, $CePO_4$, perspective c-axis projection, b vertical. Ce is in irregular seven coordination and P is tetrahedral. Both cations are in the general position.

Table 9.7.2. Monazite Cation Sites.

Site	RE	P
C.N.	7	4
Occupant	Ce	P
Point Sym.	1	1
Wyckoff Not.	4e	4e
Frac.Coord.		
x	.215	.220
y	.160	.162
z	.105	.626
Distances		
O1	2.510	1.547
O1	2.518	
O2	2.587	1.696
O2	2.509	
O3	2.723	1.495
O4	2.218	1.675
O4	2.672	
Mean	2.534	1.603
σ	.163	.098
Poly.Vol.	19.68	1.925
Q.E.		1.0681
Ang.Var.		244.
Site Energy	-1989.	-6823.

Chapter 10

HALIDES

Table 10.1.1. Halite Group Unit Cells.

End-Member	Halite	Sylvite	Villiaumite
Formula	$NaCl$	KCl	NaF
Form.Wt.	58.443	74.555	41.988
Density	2.166	1.989	2.839
Mol Volume	26.985	37.490	14.791
Z	4	4	4
Cryst.Sys.	Isometric	Isometric	Isometric
Cryst.Class	$m3m$	$m3m$	$m3m$
Space Group	$Fm3m$	$Fm3m$	$Fm3m$
Cell Parameters			
a	5.638	6.291	4.614
Vol.	179.22	248.98	98.23
Ref.	Wyckoff (1963)	Wyckoff (1963)	Wyckoff (1963)

Table 10.1.2. Halite Group Octahedral Sites.

End-Member	Halite	Sylvite	Villiaumite
C.N.	6	6	6
Occupant	Na	K	Na
Point Sym.	*m* 3*m*	*m* 3*m*	*m* 3*m*
Wyckoff Not.	4a	4a	4a
Frac.Coord.			
x	0	0	0
y	0	0	0
z	0	0	0
Distances			
O1(6)	2.819	3.145	2.307
Poly.Vol.	29.869	41.496	16.371
Q.E.	1.0000	1.0000	1.0000
Ang. Var.	0	0	0
Site Energy	-206.	-184.	-252.

Table 10.2.1. Fluorite Group Unit Cells.

End-Member	Fluorite	SrF_2	Frankdicksonite
Formula	CaF_2	SrF_2	BaF_2
Form.Wt.	78.077	125.617	175.337
Density	3.186	4.277	4.894
Mol Volume	24.509	29.373	35.824
Z	4	4	4
Cryst.Sys.	Isometric	Isometric	Isometric
Cryst.Class	$m3m$	$m3m$	$m3m$
Space Group	$Fm3m$	$Fm3m$	$Fm3m$
Cell Parameters			
a	5.460	5.7996	6.1964
Vol.	162.77	195.07	237.91
Ref.	Wyckoff (1963)	Wyckoff (1963)	Radke & Brown (1974)

Table 10.2.2. Fluorite Group Cation Sites.

End-Member	Fluorite	SrF_2	Frankdicksonite
C.N.	8	8	8
Occupant	Ca	Sr	Ba
Point Sym.	*m* 3*m*	*m* 3*m*	*m* 3*m*
Wyckoff Not.	4b	4b	4b
Frac.Coord.			
x	0	0	0
y	0	0	0
z	0	0	0
Distances			
O1(8)	2.364	2.511	2.683
Poly.Vol.	40.693	48.768	59.478
Site Energy	-850.	-800.	-749.

Chapter 11

CATION SITES LISTED BY MEAN DISTANCE

Table 11.1. Two-fold and Three-fold Sites

Mineral	Element	Site	CN	Chg	Dist	Sigma	Volume	E.E.(eV)
Cuprite	Cu	Cu	2	1.00	1.849	0.0000	0.0000	-12.8
Ag2O	Ag	Ag	2	1.00	2.044	0.0000	0.0000	-11.5
Soda Niter	N	N	3	5.00	1.218	0.0000	0.0000	-446.7
Niter	N	N	3	5.00	1.244	0.0000	0.0000	-447.3
Cerussite	C	C	3	4.00	1.270	0.0260	0.0006	-222.1
Calcite	C	C	3	4.00	1.281	0.0000	0.0000	-225.2
Aragonite	C	C	3	4.00	1.282	0.0030	0.0137	-217.5
Dolomite	C	C	3	4.00	1.284	0.0000	0.0139	-224.6
Ankerite	C	C	3	4.00	1.284	0.0000	0.0080	-224.6
Strontianite	C	C	3	4.00	1.285	0.0140	0.0048	-218.3
Magnesite	C	C	3	4.00	1.285	0.0000	0.0000	-224.6
Smithsonite	C	C	3	4.00	1.286	0.0000	0.0000	-224.6
Siderite	C	C	3	4.00	1.287	0.0000	0.0000	-224.2
Witherite	C	C	3	4.00	1.287	0.0040	0.0044	-220.1
Rhodochrosite	C	C	3	4.00	1.287	0.0000	0.0000	-224.3
Otavite	C	C	3	4.00	1.290	0.0000	0.0000	-223.8
Elbaite	B	B	3	3.00	1.361	0.0260	0.0071	-136.5
Tourmaline	B	B	3	3.00	1.374	0.0040	0.0086	-134.8
Schorl	B	B	3	3.00	1.376	0.0090	0.0092	-134.2
Claudetite	As	24e	3	3.00	1.788	0.0290	0.9345	-102.2
Claudetite	As	14e	3	3.00	1.794	0.0030	0.9469	-98.3

Table 11.2a. Four-fold Sites.

Mineral	Element	Site	CN	Chg	Dist	Sigma	Volume	Q.E.	A.V.	EE(eV)
Anhydrite	S	S	4	6.00	1.472	0.0000	1.6353	1.0012	4.82	-418.6
Celestite	S	S	4	6.00	1.474	0.0160	1.6429	1.0005	2.308	-418.5
Barite	S	S	4	6.00	1.478	0.0100	1.6549	1.0004	2.014	-417.3
Anglesite	S	S	4	6.00	1.482	0.0210	1.6686	1.0012	5.203	-415.9
Gypsum	S	S	4	6.00	1.488	0.0100	1.6891	1.0001	0.4844	-399.5
Chlorapatite	P	1	4	5.00	1.534	0.0060	1.8468	1.0017	7.097	-296.4
Apatite(OH)	P	1	4	5.00	1.535	0.0080	1.8533	1.0011	4.32	-296.3
Fluorapatite	P	1	4	5.00	1.536	0.0030	1.8566	1.0008	3.659	-296.5
Kaolinite	Si	4	4	4.00	1.569	0.0870	1.9646	1.0084	26.76	-192.6
Kirschsteinite	Si	Si	4	4.00	1.570	0.0290	1.9577	1.0102	36.83	-197.4
Tridymite	Si	9	4	4.00	1.581	0.0140	2.0263	1.0005	2.091	-197.5
Tridymite	Si	10	4	4.00	1.588	0.0180	2.0535	1.0006	2.378	-196.3
Tridymite	Si	6	4	4.00	1.588	0.0300	2.0524	1.0007	1.105	-195.5
Tridymite	Si	8	4	4.00	1.590	0.0160	2.0617	1.0002	0.65	-194.7
Dickite	Si	2	4	4.00	1.593	0.0310	2.0701	1.0012	3.358	-182.1
Tridymite	Si	12	4	4.00	1.596	0.0110	2.0864	1.0003	1.063	-193.0
Tridymite	Si	11	4	4.00	1.597	0.0260	2.0881	1.0009	1.962	-193.6
Tridymite	Si	4	4	4.00	1.597	0.0240	2.0873	1.0006	1.898	-194.9
Tridymite	Si	2	4	4.00	1.598	0.0280	2.0906	1.0007	1.819	-192.8
Tridymite	Si	5	4	4.00	1.599	0.0220	2.0974	1.0006	1.416	-193.6
Tridymite	Si	3	4	4.00	1.600	0.0100	2.0992	1.0003	1.255	-193.8
Laumontite	Si	T1	4	4.00	1.603	0.0160	2.1093	1.0016	6.191	-179.6
Monazite	P	1	4	5.00	1.603	0.0980	1.9245	1.0681	244.2	-295.9
Tridymite	Si	7	4	4.00	1.604	0.0190	2.1151	1.0012	4.333	-192.5
Tridymite	Si	1	4	4.00	1.605	0.0240	2.1192	1.0005	1.455	-192.0
Cristobalite	Si	Si	4	4.00	1.606	0.0020	2.1260	1.0005	2.142	-192.8
Anorthite	Si	0zi0	4	4.00	1.607	0.0240	2.1169	1.0047	18.35	-196.6
Na-Mordenite	Si	T2	4	3.82	1.607	0.0130	2.1257	1.0008	3.695	-194.5
Coffinite	Si	Si	4	4.00	1.607	0.0000	2.0932	1.0118	48.63	-200.6
Beryl	Si	Si	4	4.00	1.608	0.0150	2.1319	1.0006	1.689	-194.1
Marialite	Si	1	4	4.00	1.608	0.0110	2.1334	1.0003	1.605	-191.4
Coesite	Si	1	4	4.00	1.609	0.0100	2.1374	1.0003	1.209	-196.3
Low Quartz	Si	Si	4	4.00	1.609	0.0030	2.1374	1.0002	0.8828	-193.4
Kalsilite	Si	Si	4	4.00	1.609	0.0270	2.1130	1.0079	27.58	-195.4
Microcline	Si	T20	4	4.00	1.610	0.0260	2.1363	1.0025	10.04	-189.5
K-Mordenite	Si	T2	4	3.83	1.611	0.0120	2.1415	1.0011	4.811	-176.8
Microcline	Si	T2m	4	4.00	1.611	0.0240	2.1372	1.0032	13.55	-192.2
Zoisite	Si	2	4	4.00	1.612	0.0210	2.1425	1.0018	7.441	-196.4
Anorthite	Si	00i0	4	4.00	1.612	0.0200	2.1247	1.0083	32.9	-190.9
Nepheline	Si	T2	4	4.00	1.612	0.0080	2.1085	1.0131	48.96	-195.6
Coesite	Si	2	4	4.00	1.612	0.0070	2.1491	1.0001	0.4125	-193.1
Anorthite	Si	mzi0	4	4.00	1.613	0.0270	2.1373	1.0048	19.42	-195.2
Anorthite	Si	mz00	4	4.00	1.613	0.0310	2.1331	1.0061	23.23	-193.7

Table 11.2b. Four-fold Sites (continued).

Mineral	Element	Site	CN	Chg	Dist	Sigma	Volume	Q.E.	A.V.	EE(eV)
Epidote 1	Si	2	4	4.00	1.614	0.0190	2.1551	1.0009	3.686	-197.0
Clinoptilolite	Si	T5	4	3.74	1.614	0.0030	2.1538	1.0013	5.336	-174.3
Microcline	Si	T1m	4	4.00	1.614	0.0170	2.1545	1.0014	4.955	-189.4
Epidote 2	Si	2	4	4.00	1.614	0.0160	2.1551	1.0012	4.969	-198.0
Laumontite	Si	T3	4	4.00	1.614	0.0340	2.1385	1.0061	25.07	-167.5
Low Albite	Si	T20	4	4.00	1.614	0.0160	2.1509	1.0022	9.567	-189.8
Cordierite	Si	T21	4	4.00	1.614	0.0260	2.1553	1.0010	4.392	-193.0
Clinozoisite	Si	2	4	4.00	1.615	0.0150	2.1572	1.0014	6.022	-196.9
Anorthite	Si	0000	4	4.00	1.615	0.0260	2.1246	1.0121	47.87	-191.4
Clinoptilolite	Si	T4	4	3.74	1.615	0.0040	2.1587	1.0010	4.152	-172.6
Anorthite	Si	m000	4	4.00	1.615	0.0280	2.1526	1.0027	10.84	-194.0
Anorthite	Si	m0i0	4	4.00	1.615	0.0220	2.1586	1.0015	6.394	-192.8
Anorthite	Si	0z00	4	4.00	1.616	0.0140	2.1459	1.0065	25.61	-192.2
Low Albite	Si	T2m	4	4.00	1.616	0.0220	2.1578	1.0021	8.989	-190.5
Bustamite	Si	1	4	4.00	1.616	0.0250	2.1510	1.0050	21.6	-192.4
Glaucophane	Si	T1	4	4.00	1.616	0.0040	2.1660	1.0002	0.8047	-190.2
Nepheline	Si	T3	4	4.00	1.616	0.0280	2.1619	1.0012	3.72	-193.2
Gedrite	Si	A	4	3.78	1.616	0.0260	2.1396	1.0087	33.4	-158.2
Elbaite	Si	Si	4	4.00	1.616	0.0100	2.1613	1.0019	7.88	-190.0
Bustamite	Si	2	4	4.00	1.617	0.0260	2.1540	1.0050	21.8	-192.1
Thomsonite	Si	1	4	4.00	1.617	0.0010	2.1689	1.0006	2.573	-186.4
Rhodonite	Si	3	4	4.00	1.617	0.0180	2.1400	1.0090	37.3	-191.9
Pyrophyllite	Si	1	4	4.00	1.617	0.0120	2.1684	1.0005	2.039	-190.5
Nacrite	Si	1	4	4.00	1.617	0.0200	2.1550	1.0044	17.34	-183.3
Cordierite	Si	T23	4	4.00	1.617	0.0270	2.1662	1.0015	6.272	-192.1
Pyroxmangite	Si	6	4	4.00	1.617	0.0140	2.1310	1.0120	47.1	-191.9
Pyrophyllite	Si	2	4	4.00	1.618	0.0120	2.1714	1.0004	1.684	-190.9
Clinoptilolite	Si	T3	4	3.74	1.618	0.0110	2.1692	1.0010	3.681	-168.9
Akermanite	Si	Si	4	4.00	1.619	0.0230	2.1450	1.0110	47.1	-189.9
Spodumene	Si	Si	4	4.00	1.619	0.0220	2.1640	1.0050	18.4	-193.4
K-Mordenite	Si	T1	4	3.83	1.619	0.0090	2.1735	1.0013	5.27	-179.5
K-Heulandite	Si	T4	4	3.74	1.619	0.0100	2.1738	1.0018	7.394	-171.1
Kaolinite	Si	1	4	4.00	1.619	0.0650	2.1335	1.0151	40.47	-192.9
Thomsonite	Si	3	4	4.00	1.619	0.0060	2.1688	1.0027	10.5	-187.9
Gedrite	Si	A	4	3.78	1.620	0.0130	2.1833	1.0002	0.9	-169.0
Nacrite	Si	2	4	4.00	1.620	0.0130	2.1694	1.0037	15.42	-183.0
Rhodonite	Si	2	4	4.00	1.620	0.0200	2.1680	1.0050	20.5	-191.8
Heulandite	Si	T4	4	3.74	1.620	0.0090	2.1799	1.0009	4.158	-170.9
Wollastonite	Si	1	4	4.00	1.620	0.0320	2.1640	1.0060	26.2	-192.3
Schorl	Si	Si	4	4.00	1.620	0.0170	2.1746	1.0019	7.647	-190.4
Wollastonite	Si	2	4	4.00	1.620	0.0280	2.1640	1.0050	22.2	-191.2
Tremolite	Si	T1	4	4.00	1.621	0.0140	2.1830	1.0012	5.127	-189.3
Hafnon	Si	Si	4	4.00	1.621	0.0010	2.1090	1.0242	99.65	-196.0

Table 11.2c. Four-fold Sites (continued).

Mineral	Element	Site	CN	Chg	Dist	Sigma	Volume	Q.E.	A.V.	EE(eV)
Na-melilite	Si	T3	4	4.00	1.622	0.0310	2.1630	1.0080	36.6	-188.3
Gedrite	Si	B	4	3.78	1.622	0.0110	2.1893	1.0001	0.3375	-206.3
Margarite	Si	T11	4	4.00	1.622	0.0070	2.1863	1.0006	2.355	-188.8
Talc	Si	2	4	4.00	1.622	0.0010	2.1914	1.0000	0.2594	-189.9
Na-Mordenite	Si	T1	4	3.82	1.622	0.0130	2.1824	1.0023	9.225	-195.1
Clinoferrosilite	Si	A	4	4.00	1.622	0.0270	2.1730	1.0060	24.6	-192.1
Allanite	Si	2	4	4.00	1.623	0.0140	2.1918	1.0011	4.133	-195.4
Rhodonite	Si	4	4	4.00	1.623	0.0250	2.1810	1.0040	14.5	-191.1
Talc	Si	1	4	4.00	1.623	0.0020	2.1936	1.0000	0.2484	-189.8
Heulandite	Si	T3	4	3.74	1.623	0.0060	2.1899	1.0015	5.925	-168.2
Rhodonite	Si	1	4	4.00	1.623	0.0260	2.1550	1.0115	51.6	-191.8
Zircon	Si	Si	4	4.00	1.623	0.0000	2.1179	1.0237	97.75	-195.7
Willemite	Si	Si	4	4.00	1.623	0.0110	2.1914	1.0009	3.291	-192.8
Low Albite	Si	T1m	4	4.00	1.624	0.0240	2.1847	1.0044	15.78	-189.2
Na-Mordenite	Si	T4	4	3.82	1.624	0.0110	2.1914	1.0024	10.62	-186.5
Gedrite	Si	A	4	3.78	1.624	0.0230	2.1745	1.0076	30.45	-155.5
Clinoptilolite	Si	T1	4	3.74	1.624	0.0050	2.1952	1.0010	4.488	-167.2
Epidote 1	Si	1	4	4.00	1.624	0.0410	2.1903	1.0033	9.867	-188.6
Clinozoisite	Si	1	4	4.00	1.625	0.0410	2.1904	1.0034	10.46	-190.0
Heulandite	Si	T1	4	3.74	1.625	0.0040	2.1971	1.0009	4.016	-169.3
Epidote 2	Si	1	4	4.00	1.625	0.0430	2.1914	1.0035	10.49	-189.7
Tourmaline	Si	Al	4	4.00	1.625	0.0160	2.1954	1.0024	9.798	-186.2
CoOrthopyrox	Si	A	4	4.00	1.625	0.0240	2.1800	1.0080	30.7	-192.1
Jadeite	Si	Si	4	4.00	1.625	0.0210	2.1820	1.0060	23.1	-192.6
Thomsonite	Si	2	4	3.92	1.626	0.0070	2.2000	1.0015	6.247	-181.7
Ureyite	Si	Si	4	4.00	1.626	0.0260	2.1960	1.0040	15.9	-192.8
Pyroxmangite	Si	5	4	4.00	1.626	0.0330	2.1770	1.0090	35.5	-191.8
Cordierite	Si	T16	4	4.00	1.626	0.0000	2.1333	1.0226	88.63	-187.7
Orthoferrosilite	Si	A	4	4.00	1.626	0.0220	2.1810	1.0080	31.1	-192.0
Sillimanite	Si	Si	4	4.00	1.626	0.0350	2.2029	1.0013	3.43	-192.0
Clinohumite	Si	1	4	4.00	1.626	0.0040	2.1750	1.0094	39.65	-186.5
Cummingtonite	Si	T1	4	4.00	1.627	0.0110	2.2117	1.0002	0.6438	-201.9
Acmite	Si	Si	4	4.00	1.627	0.0210	2.2010	1.0030	13.9	-192.7
Humite	Si	2	4	4.00	1.627	0.0110	2.1800	1.0096	42.47	-188.5
Rhodonite	Si	5	4	4.00	1.627	0.0240	2.1950	1.0040	19.1	-189.3
Heulandite	Si	T5	4	3.74	1.627	0.0040	2.2053	1.0010	4.159	-173.2
Tremolite	Si	T1	4	4.00	1.627	0.0140	2.2058	1.0014	6.178	-186.2
Lepidolite	Si	T2	4	3.84	1.627	0.0120	2.2059	1.0019	8.52	-174.6
Vesuvianite	Si	3	4	4.00	1.628	0.0200	2.2113	1.0010	4.284	-184.8
Orthoenstatite	Si	A	4	4.00	1.628	0.0350	2.1820	1.0099	39.8	-191.1
Cummingtonite	Si	T2	4	4.00	1.628	0.0180	2.2017	1.0040	15.66	-176.5
K-Heulandite	Si	T3	4	3.74	1.628	0.0050	2.2058	1.0022	8.652	-171.3
Almandine	Si	Z	4	4.00	1.628	0.0010	2.1717	1.0134	55.2	-191.4

Table 11.2d. Four-fold Sites (continued).

Mineral	Element	Site	CN	Chg	Dist	Sigma	Volume	Q.E.	A.V.	EE(eV)
Lepidolite	Si	T2	4	3.94	1.628	0.0270	2.2039	1.0038	14.31	-189.5
Pyroxmangite	Si	4	4	4.00	1.629	0.0110	2.2030	1.0050	19.2	-191.8
Heulandite	Si	T4	4	3.74	1.629	0.0140	2.2144	1.0010	4.355	-170.9
Orthoclase	Si	T2	4	4.00	1.629	0.0090	2.2108	1.0017	7.313	-188.1
Pyroxmangite	Si	1	4	4.00	1.629	0.0260	2.2090	1.0040	13.3	-190.4
Humite	Si	1	4	4.00	1.629	0.0090	2.1878	1.0090	38.18	-186.2
K-Mordenite	Si	T4	4	3.83	1.629	0.0020	2.2124	1.0020	8.173	-174.2
Norbergite	Si	Si	4	4.00	1.630	0.0120	2.1932	1.0093	41.26	-187.4
Pyroxmangite	Si	2	4	4.00	1.630	0.0260	2.2110	1.0040	16.5	-189.1
Lepidolite	Si	T1	4	3.84	1.630	0.0180	2.2168	1.0017	7.945	-174.3
Heulandite	Si	T3	4	3.74	1.630	0.0060	2.2188	1.0009	3.862	-166.2
Allanite	Si	1	4	4.00	1.630	0.0330	2.2165	1.0027	8.911	-185.9
Phenacite	Si	Si	4	4.00	1.631	0.0020	2.2220	1.0009	3.809	-188.2
Lizardite	Si	Si	4	4.00	1.631	0.0070	2.2260	1.0000	0.2547	-185.2
Andalusite	Si	Si	4	4.00	1.631	0.0110	2.2109	1.0043	16.42	-191.0
Laumontite	Si	T4	4	3.85	1.631	0.0150	2.2197	1.0027	10.6	-154.3
K-Heulandite	Si	T1	4	3.74	1.631	0.0120	2.2199	1.0016	7.098	-171.1
Lepidolite	Si	T	4	3.84	1.631	0.0180	2.2171	1.0025	10.48	-173.8
Bustamite	Si	3	4	4.00	1.632	0.0340	2.1850	1.0142	63.87	-190.0
Pyroxmangite	Si	7	4	4.00	1.632	0.0370	2.1950	1.0110	48.8	-190.1
Allanite	Si	3	4	4.00	1.632	0.0320	2.2014	1.0086	33.52	-196.0
Gedrite	Si	B	4	3.78	1.633	0.0200	2.2207	1.0046	18.35	-163.1
Tremolite	Si	T2	4	4.00	1.633	0.0380	2.2186	1.0050	19.8	-183.1
Margarite	Si	T2	4	4.00	1.633	0.0060	2.2306	1.0009	3.617	-185.9
Lawsonite	Si	Si	4	4.00	1.633	0.0200	2.2193	1.0049	19.61	-207.3
Chondrodite	Si	Si	4	4.00	1.633	0.0120	2.2023	1.0102	45.17	-186.8
MnClinopyrox	Si	A	4	4.00	1.633	0.0290	2.2160	1.0060	23.2	-191.3
Pyroxmangite	Si	3	4	4.00	1.633	0.0330	2.2210	1.0040	17	-190.3
Melilite	Si	T3	4	3.50	1.633	0.0280	2.2010	1.0110	49.9	-187.3
Rb Feldspar	Si	T2	4	4.00	1.634	0.0080	2.2336	1.0013	5.122	-187.1
K-Mordenite	Si	T3	4	3.83	1.634	0.0270	2.2371	1.0013	5.031	-172.7
Wollastonite	Si	3	4	4.00	1.634	0.0340	2.1990	1.0130	57.9	-190.5
Heulandite	Si	T1	4	3.74	1.635	0.0090	2.2412	1.0007	3.017	-169.2
Thorite	Si	Si	4	4.00	1.635	0.0010	2.2054	1.0109	44.78	-197.1
Hedenbergite	Si	Si	4	4.00	1.635	0.0490	2.2240	1.0060	24.9	-190.8
Pyrope	Si	Z	4	4.00	1.635	0.0000	2.1919	1.0150	61.58	-190.0
Diopside	Si	Si	4	4.00	1.635	0.0490	2.2210	1.0070	28.6	-190.6
Clinoferrosilite	Si	B	4	4.00	1.635	0.0340	2.2280	1.0050	19.3	-189.4
Zoisite	Si	1	4	4.00	1.635	0.0330	2.2258	1.0049	17.45	-189.5
Celsian	Si	0z00	4	4.00	1.635	0.0120	2.2277	1.0043	15.99	-184.2
Kyanite	Si	1	4	4.00	1.636	0.0110	2.2412	1.0012	4.83	-192.7
Fayalite	Si	Si	4	4.00	1.636	0.0120	2.2196	1.0085	36.68	-188.6
Kyanite	Si	2	4	4.00	1.636	0.0100	2.2428	1.0018	7.145	-193.4

Table 11.2e. Four-fold Sites (continued).

Mineral	Element	Site	CN	Chg	Dist	Sigma	Volume	Q.E.	A.V.	EE(eV)
Bromellite	Be	Be	4	2.00	1.636	0.0110	2.2475	1.0005	1.475	-57.8
Pargasite	Si	T2	4	3.91	1.636	0.0230	2.2299	1.0047	19.31	-186.2
Rb Feldspar	Al	T1	4	3.50	1.636	0.0050	2.2409	1.0016	6.086	-153.2
Forsterite	Si	Si	4	4.00	1.636	0.0170	2.2086	1.0112	49.43	-187.3
Co-Olivine	Si	Si	4	4.00	1.636	0.0190	2.2161	1.0100	44.31	-187.6
Heulandite	Si	T5	4	3.74	1.636	0.0150	2.2429	1.0016	6.206	-173.3
Spessartine	Si	Z	4	4.00	1.636	0.0000	2.2064	1.0117	48	-190.8
Tremolite	Si	T2	4	4.00	1.636	0.0420	2.2289	1.0054	21.66	-194.6
Celsian	Si	0000	4	4.00	1.637	0.0080	2.2263	1.0079	30.48	-180.2
Monticellite	Si	Si	4	4.00	1.637	0.0170	2.2202	1.0092	40.59	-187.9
Sanidine	Si	T2	4	3.75	1.637	0.0040	2.2455	1.0019	7.753	-169.2
MnClinopyrox	Si	B	4	4.00	1.637	0.0360	2.2400	1.0040	14.7	-189.8
Clinohumite	Si	2	4	4.00	1.638	0.0140	2.2192	1.0109	48.56	-187.9
Orthoferrosilite	Si	B	4	4.00	1.638	0.0320	2.2410	1.0050	17.3	-188.8
CoOrthopyrox	Si	B	4	4.00	1.638	0.0340	2.2410	1.0050	17.6	-188.5
Liebenbergite	Si	Si	4	4.00	1.638	0.0160	2.2177	1.0118	52.26	-186.7
Amesite	Si	T1	4	4.00	1.639	0.0360	2.2568	1.0008	0.7141	-181.6
Na-Mordenite	Si	T3	4	3.82	1.639	0.0140	2.2498	1.0028	11.95	-189.1
Tephroite	Si	Si	4	4.00	1.639	0.0160	2.2322	1.0082	36.11	-188.6
Dickite	Si	3	4	4.00	1.639	0.0730	2.2426	1.0064	15.58	-173.8
Analcime	Si	T1	4	3.73	1.639	0.0060	2.2517	1.0026	10.59	-167.5
Glaucophane	Si	T2	4	4.00	1.640	0.0340	2.2469	1.0047	17.58	-182.5
Vesuvianite	Si	1	4	4.00	1.640	0.0000	2.2454	1.0050	20.61	-168.8
Zoisite	Si	3	4	4.00	1.640	0.0260	2.2273	1.0114	44.29	-190.8
Orthoenstatite	Si	B	4	4.00	1.640	0.0440	2.2470	1.0050	19.4	-188.7
Harmotome	Si	T3	4	3.71	1.640	0.0090	2.2598	1.0017	6.68	-165.7
Clinoenstatite	Si	A	4	4.00	1.640	0.0300	2.2490	1.0044	16.7	-189.8
K-Heulandite	Si	T5	4	3.74	1.640	0.0130	2.2591	1.0018	7.308	-170.6
Harmotome	Si	T2	4	3.71	1.640	0.0070	2.2579	1.0011	4.738	-166.8
Epidote 1	Si	3	4	4.00	1.640	0.0190	2.2493	1.0046	18.39	-192.0
Kaolinite	Si	2	4	4.00	1.641	0.0870	2.2492	1.0072	24.59	-187.0
Epidote 2	Si	3	4	4.00	1.641	0.0200	2.2508	1.0055	22.21	-190.4
Malayaite	Si	Si	4	4.00	1.641	0.0090	2.2526	1.0044	17.36	-192.9
High Albite	Si	T20	4	3.75	1.641	0.0110	2.2626	1.0017	7.022	-169.6
Co beta Spinel	Si	Si	4	4.00	1.641	0.0380	2.2575	1.0029	11.64	-189.3
Staurolite	Si	Si	4	4.00	1.641	0.0080	2.2660	1.0004	1.658	-189.1
F-Phlogopite	Si	T	4	4.00	1.642	0.0060	2.2695	1.0002	0.7344	-181.9
High Albite	Si	T1m	4	3.75	1.642	0.0120	2.2667	1.0019	7.423	-169.8
Zinnwaldite	Si	T11	4	3.50	1.642	0.0030	2.2697	1.0009	3.714	-153.3
Uvarovite	Si	Z	4	4.00	1.643	0.0000	2.2572	1.0058	23.9	-192.0
High Albite	Si	T2m	4	3.75	1.643	0.0100	2.2728	1.0007	3.031	-168.3
Andradite	Si	Z	4	4.00	1.643	0.0010	2.2503	1.0071	28.92	-192.2
Lepidolite	Si	T1	4	3.85	1.643	0.0230	2.2657	1.0037	14.19	-180.6

Table 11.2f. Four-fold Sites (continued).

Mineral	Element	Site	CN	Chg	Dist	Sigma	Volume	Q.E.	A.V.	EE(eV)
Leucite	Si	1	4	3.67	1.644	0.0070	2.2711	1.0022	8.731	-165.1
Clinozoisite	Si	3	4	4.00	1.644	0.0180	2.2606	1.0060	23.87	-190.6
Muscovite	Si	T2	4	3.75	1.644	0.0030	2.2782	1.0010	4.111	-168.0
Titanite	Si	Si	4	4.00	1.645	0.0030	2.2729	1.0032	12.12	-191.7
Grossular	Si	Z	4	4.00	1.645	0.0000	2.2612	1.0073	29.82	-190.9
Clinoptilolite	Si	T2	4	3.74	1.645	0.0110	2.2812	1.0012	4.469	-166.2
Muscovite	Si	T1	4	3.75	1.645	0.0070	2.2785	1.0012	4.973	-168.0
Harmotome	Si	T4	4	3.71	1.645	0.0050	2.2785	1.0018	6.898	-165.6
Phenacite	Be	1	4	2.00	1.645	0.0090	2.2803	1.0014	5.383	-59.8
Ca-Olivine	Si	Si	4	4.00	1.646	0.0090	2.2663	1.0062	27.19	-189.8
gammaCo2SiO4	Si	A	4	4.00	1.646	0.0000	2.2900	1.0000	0.2219	-193.5
Phenacite	Be	2	4	2.00	1.646	0.0110	2.2829	1.0017	7.191	-60.6
Vesuvianite	Si	2	4	4.00	1.646	0.0260	2.2636	1.0074	30.16	-187.4
Mg beta spinel	Si	Si	4	4.00	1.651	0.0340	2.2972	1.0037	14.72	-187.4
Amesite	Si	T22	4	4.00	1.649	0.0320	2.2993	1.0007	1.359	-180.9
Zinnwaldite	Si	T1	4	3.50	1.649	0.0040	2.2986	1.0009	3.578	-149.9
Phlogopite	Si	T	4	4.00	1.649	0.0020	2.3009	1.0003	1.498	-186.4
Sanidine	Si	T1	4	3.75	1.649	0.0080	2.2935	1.0026	10.07	-167.6
High Albite	Al	T10	4	3.75	1.649	0.0050	2.2869	1.0040	15.97	-166.5
Gedrite	Si	B	4	3.78	1.649	0.0170	2.2861	1.0041	15.54	-161.3
Marialite	Si	3	4	3.71	1.650	0.0150	2.2920	1.0043	17.46	-161.3
Gedrite	Si	A	4	3.78	1.650	0.0170	2.3055	1.0006	2.336	-163.1
Phillipsite	Si	T1	4	3.67	1.651	0.0080	2.2997	1.0029	11.94	-160.8
Paragonite	Si	T2	4	4.00	1.652	0.0070	2.3119	1.0007	3.014	-321.7
gammaFe2SiO4	Si	A	4	4.00	1.652	0.0010	2.3116	1.0000	0.2219	-193.4
Meionite	Si	1	4	3.69	1.653	0.0250	2.3091	1.0022	9.617	-164.9
Harmotome	Si	T1	4	3.71	1.653	0.0110	2.3083	1.0029	12	-163.8
Paragonite	Si	T1	4	4.00	1.653	0.0040	2.3138	1.0007	2.855	-327.4
Phillipsite	Si	T2	4	3.67	1.654	0.0080	2.3185	1.0016	6.617	-161.3
Beryl	Be	Be	4	2.00	1.654	0.0010	2.0268	1.0950	331.7	-58.5
gammaNi2SiO4	Si	A	4	4.00	1.654	0.0010	2.3213	1.0000	0.2156	-191.7
Phillipsite	Si	T4	4	3.67	1.654	0.0100	2.3154	1.0015	6.089	-163.0
Orthoclase	Al	T1	4	3.50	1.655	0.0110	2.3206	1.0022	8.716	-150.5
GammaMg2SiO4	Si	A	4	4.00	1.655	0.0000	2.3276	1.0000	0.2219	-191.6
Gedrite	Si	B	4	3.78	1.656	0.0090	2.3311	1.0001	0.6938	-197.5
K-Heulandite	Si	T2	4	3.74	1.657	0.0130	2.3322	1.0012	3.892	-166.8
Leucite	Si	3	4	3.67	1.657	0.0140	2.3307	1.0006	2.275	-159.1
Leucite	Si	2	4	3.67	1.657	0.0080	2.3291	1.0014	5.392	-161.2
Heulandite	Si	T2	4	3.74	1.657	0.0120	2.3270	1.0028	10.58	-164.2
Heulandite	Si	T2	4	3.74	1.658	0.0130	2.3325	1.0019	6.828	-163.9
Kaolinite	Si	3	4	4.00	1.659	0.0820	2.3251	1.0076	20.55	-177.8
Annite	Si	T	4	3.75	1.660	0.0060	2.3451	1.0003	1.155	-166.1
Clinoenstatite	Si	B	4	4.00	1.662	0.0680	2.3280	1.0040	34.1	-185.5

Table 11.2g. Four-fold Sites (continued).

Mineral	Element	Site	CN	Chg	Dist	Sigma	Volume	Q.E.	A.V.	EE(eV)
Analcime	Si	T2	4	3.55	1.663	0.0080	2.3457	1.0036	14.6	-152.5
Phillipsite	Si	T3	4	3.67	1.664	0.0110	2.3542	1.0028	11.39	-160.5
Pargasite	Si	T1	4	3.62	1.675	0.0150	2.4073	1.0016	6.769	-153.5
Phlogopite-N	Si	T	4	4.00	1.678	0.0480	2.2094	1.0642	216.1	-182.6
Meionite	Si	3	4	3.51	1.678	0.0040	2.4030	1.0063	24.74	-143.2
Meionite	Si	2	4	3.49	1.681	0.0080	2.4135	1.0061	23.88	-141.5
Ca-Tschermakite	Si	T	4	3.50	1.685	0.0150	2.4250	1.0090	35.8	-149.0
Gehlenite	Al	T3	4	3.50	1.691	0.0190	2.4310	1.0140	61.5	-144.8
Marialite	Si	2	4	3.41	1.693	0.0050	2.4742	1.0041	16.49	-136.1
Celsian	Al	0000	4	3.00	1.712	0.0150	2.5468	1.0069	24.33	-109.5
Laumontite	Al	T2	4	3.32	1.712	0.0720	2.5325	1.0122	44.24	-117.9
Nepheline	Al	T1	4	3.00	1.713	0.0050	2.5242	1.0147	54.63	-115.3
Celsian	Al	0z00	4	3.00	1.719	0.0140	2.5686	1.0099	38.38	-106.5
Amesite	Al	T11	4	3.00	1.725	0.0100	2.6306	1.0005	1.917	-106.1
Amesite	Al	T2	4	3.00	1.729	0.0290	2.6515	1.0006	1.777	-105.3
Nepheline	Al	T4	4	3.00	1.734	0.0240	2.6711	1.0018	5.8	-110.9
Microcline	Al	T10	4	3.00	1.741	0.0030	2.6936	1.0032	12.44	-111.7
Thomsonite	Al	3	4	3.00	1.741	0.0070	2.6878	1.0050	19.38	-106.9
Anorthite	Al	0000	4	3.00	1.741	0.0310	2.6944	1.0039	14.98	-112.2
Thomsonite	Al	1	4	3.00	1.741	0.0020	2.7029	1.0012	5.031	-105.0
Cordierite	Al	T26	4	3.00	1.742	0.0360	2.7053	1.0025	10.63	-112.6
Thomsonite	Al	2	4	3.00	1.743	0.0110	2.7051	1.0027	11.04	-106.1
Anorthite	Al	m0i0	4	3.00	1.744	0.0310	2.6903	1.0079	30.27	-111.1
Anorthite	Al	mz00	4	3.00	1.744	0.0190	2.6990	1.0057	22.06	-108.8
Kalsilite	Al	Al	4	3.00	1.744	0.0300	2.6984	1.0061	23.96	-111.2
Anorthite	Al	mzi0	4	3.00	1.745	0.0280	2.7157	1.0024	9.041	-110.9
Anorthite	Al	0z00	4	3.00	1.747	0.0290	2.6672	1.0171	64.56	-107.7
Margarite	Al	T1	4	3.00	1.747	0.0210	2.7288	1.0017	7.347	-105.9
Anorthite	Al	m000	4	3.00	1.749	0.0350	2.7185	1.0075	29.84	-111.2
Low Albite	Al	T10	4	3.00	1.750	0.0190	2.7004	1.0128	51.26	-109.7
Anorthite	Al	00i0	4	3.00	1.750	0.0170	2.6904	1.0146	54.13	-108.5
Anorthite	Al	0zi0	4	3.00	1.755	0.0200	2.6666	1.0269	99.96	-107.9
Cordierite	Al	T11	4	3.00	1.758	0.0010	2.5809	1.0530	195.7	-104.9
Na-melilite	Al	T1	4	3.00	1.762	0.0000	2.7880	1.0040	16.8	-105.9
Sillimanite	Al	2	4	3.00	1.764	0.0410	2.7910	1.0062	20.48	-109.6
Gehlenite	Al	T1	4	3.00	1.785	0.0000	2.9160	1.0010	3.6	-105.4
Scheelite	W	W	4	6.00	1.785	0.0000	2.9099	1.0024	9.652	-356.7
Melilite	Mg	T1	4	2.00	1.876	0.0000	3.3820	1.0010	4	-74.6
Magnetite	Fe	A	4	3.00	1.887	0.0010	3.4490	1.0000	0.2219	-103.4
Magnesioferrite	Fe	A	4	2.90	1.911	0.0010	3.5835	1.0000	0.2156	-95.9
Akermanite	Mg	Mg	4	2.00	1.915	0.0000	3.5990	1.0010	5.8	-51.4
Spinel	Mg	A	4	2.07	1.924	0.0010	3.6532	1.0000	0.2156	-55.0
Hercynite	Fe	A	4	2.00	1.954	0.0000	3.8267	1.0000	0.2156	-51.1

Table 11.2h. Four-fold Sites (continued).

Mineral	Element	Site	CN	Chg	Dist	Sigma	Volume	Q.E.	A.V.	EE(eV)
Tenorite	Cu	4c	4	2.00	1.956	0.0060	0.0000	planar	912.7	-48.3
Willemite	Zn	1	4	2.00	1.958	0.0050	3.8209	1.0054	21.43	-48.7
Magnesiochromite	Mg	A	4	2.00	1.966	0.0000	3.8990	1.0000	0.2172	-51.6
Zincite	Zn	Zn	4	2.00	1.974	0.0090	3.9421	1.0006	2.58	-47.9
Willemite	Zn	2	4	2.00	1.976	0.0220	3.9342	1.0048	19.63	-48.1
Paratellurite	Te	Te	4	4.00	2.003	0.0970	2.5084	1.3959	651	-164.1
Chromite	Fe	A	4	2.00	2.006	0.0000	4.1409	1.0000	0.2156	-49.8
Ulvoespinel	Fe	A	4	2.00	2.011	0.0000	4.1720	1.0000	0.2188	-50.5
Jacobsite	Mn	A	4	2.15	2.012	0.0000	4.1812	1.0000	0.2203	-56.1
Tellurite	Te	Te	4	4.00	2.018	0.1440	2.4944	1.4240	544.5	-163.0
Cummingtonite	Mg	M4	4	2.00	2.108	0.0780	3.1910	1.3161	755.4	-34.5

Table 11.3. Five-fold Sites.

Mineral	Element	Site	CN	Chg	Dist	Sigma	Volume	E.E.(eV)
Andalusite	Al	2	5	3.00	1.836	0.0360	5.1527	-111.4
Vesuvianite	B	B	5	1.00	2.107	0.0620	6.6041	-10.0
Valentinite	Sb	8e	5	3.00	2.231	0.3110	5.4679	-94.7
Montroydite	Hg	4c	5	2.00	2.508	0.4450	10.7091	-42.1

Table 11.4a. Six-fold Sites.

Mineral	Element	Site	CN	Chg	Dist	Sigma	Volume	Q.E.	A.V.	EE(eV)
Stishovite	Si	Si	6	4.00	1.775	0.0270	7.3652	1.0080	27.15	-197.3
MgSiO3 Perov	Si	Si	6	4.00	1.793	0.0080	7.6808	1.0005	1.614	-195.9
MgSiO3 Ilmenite	Si	Si	6	4.00	1.799	0.0340	7.5920	1.0152	52.76	-188.8
Dickite	Al	3	6	3.00	1.832	0.1230	7.8466	1.0335	98.59	-127.3
Clinozoisite	Al	2	6	3.00	1.878	0.0350	8.7734	1.0045	14.04	-114.2
Epidote 2	Al	M2	6	3.00	1.883	0.0340	8.8471	1.0042	13.09	-116.4
Epidote 1	Al	M2	6	3.00	1.883	0.0340	8.8532	1.0045	14.01	-113.4
Zinnwaldite	Al	M2	6	3.00	1.885	0.0180	8.8957	1.0025	8.515	-97.4
Kaolinite	Al	2	6	3.00	1.885	0.0790	8.6758	1.0205	64.43	-91.8
Pyrope	Al	Y	6	3.00	1.887	0.0010	8.9373	1.0014	4.928	-115.6
Pyrolusite	Mn	Mn	6	4.00	1.887	0.0070	8.8468	1.0079	27.99	-186.0
Staurolite	Fe	Fe	6	2.36	1.888	0.1900	7.3801	1.1482	386	-60.5
Vesuvianite	Al	Al	6	3.00	1.888	0.0400	8.8929	1.0068	21.51	-117.4
Zoisite	Al	12	6	3.00	1.888	0.0500	8.8993	1.0066	20.05	-111.6
Ramsdellite	Mn	Mn	6	4.00	1.891	0.0480	8.7980	1.0169	54.1	-179.0
Almandine	Al	Y	6	3.00	1.896	0.0010	9.0862	1.0004	1.381	-115.2
Kyanite	Al	4	6	3.00	1.896	0.0650	8.9209	1.0139	42.49	-109.8
Spessartine	Al	Y	6	3.00	1.901	0.0010	9.1547	1.0001	0.2997	-115.3
Kyanite	Al	1	6	3.00	1.902	0.0620	8.9773	1.0155	47.7	-109.8
Gibbsite	Al	1	6	3.00	1.902	0.0350	9.0188	1.0120	42.21	-104.4
Margarite	Al	M2	6	3.00	1.903	0.0340	9.0058	1.0136	47.93	-116.7
Allanite	Al	M2	6	3.00	1.904	0.0210	9.1103	1.0064	22.12	-114.5
Boehmite	Al	Al	6	3.00	1.905	0.0310	9.0000	1.0164	52.07	-95.6
Elbaite	Al	Mg	6	3.00	1.905	0.0420	9.0308	1.0140	46.89	-107.5
Staurolite	Al	2	6	3.00	1.905	0.0280	9.0314	1.0139	46.26	-113.6
Gibbsite	Al	2	6	3.00	1.905	0.0330	9.0610	1.0120	42.53	-103.9
Clinozoisite	Al	1	6	3.00	1.906	0.0430	9.1457	1.0065	19.65	-105.6
Beryl	Al	Al	6	3.00	1.906	0.0010	8.9385	1.0218	75.53	-90.5
Epidote 2	Al	M1	6	3.00	1.907	0.0470	9.1674	1.0064	18.85	-105.3
Paragonite	Al	M2	6	2.00	1.908	0.0160	9.0395	1.0162	58.98	-110.6
Kaolinite	Al	4	6	3.00	1.910	0.1340	8.9201	1.0313	92.96	-94.6
Staurolite	Al	1A	6	3.00	1.911	0.0210	9.1273	1.0133	45.47	-102.6
Pyrophyllite	Al	Al	6	3.00	1.912	0.0180	9.0689	1.0183	66.43	-105.7
Sillimanite	Al	1	6	3.00	1.912	0.0390	9.1748	1.0109	36.36	-111.6
Dickite	Al	2	6	3.00	1.912	0.1140	8.7597	1.0450	140.6	-118.8
Kyanite	Al	2	6	3.00	1.913	0.0230	9.1361	1.0141	50.21	-111.2
Corundum	Al	Al	6	3.00	1.913	0.0620	9.0660	1.0200	66.59	-109.7
Lawsonite	Al	Al	6	3.00	1.913	0.0420	9.1900	1.0112	36.74	-116.2
Epidote 1	Al	M1	6	3.00	1.913	0.0540	9.2522	1.0065	17.94	-104.5
Staurolite	Al	1B	6	3.00	1.914	0.0220	9.1685	1.0132	45.32	-102.6
Nacrite	Al	1	6	3.00	1.915	0.0750	9.1086	1.0200	66.2	-110.2
Margarite	Al	M3	6	3.00	1.915	0.0540	9.1512	1.0160	54.95	-115.6
Diaspore	Al	Al	6	3.00	1.915	0.0670	9.0996	1.0205	63.98	-107.7

Table 11.4b. Six-fold Sites (continued).

Mineral	Element	Site	CN	Chg	Dist	Sigma	Volume	Q.E.	A.V.	EE(eV)
Kaolinite	Al	1	6	3.00	1.916	0.1140	9.0221	1.0295	91.94	-95.6
Kyanite	Al	3	6	3.00	1.918	0.0500	9.1639	1.0180	57.01	-110.3
Nacrite	Al	2	6	3.00	1.920	0.0340	9.1924	1.0178	63.21	-110.0
Spodumene	Al	M1	6	3.00	1.921	0.0820	9.2620	1.0150	44.4	-112.1
Schorl	Al	MgZ	6	3.00	1.922	0.0420	9.2968	1.0127	42.47	-105.7
Grossular	Al	Y	6	3.00	1.924	0.0000	9.4905	1.0007	2.331	-114.5
Spinel	Al	B	6	2.96	1.926	0.0010	9.3710	1.0108	40.79	-104.4
Perovskite	Ti	Ti	6	4.00	1.926	0.0020	9.4919	1.0019	6.682	-184.6
Jadeite	Al	M1	6	3.00	1.929	0.0640	9.3730	1.0150	47.8	-112.2
Tourmaline	Al	Z	6	3.00	1.929	0.0400	9.3790	1.0141	47.9	-105.1
Muscovite	Al	M	6	3.00	1.930	0.0120	9.3550	1.0162	59.02	-104.2
Glaucophane	Al	M2	6	3.00	1.930	0.0870	9.4346	1.0121	35.84	-106.5
Andalusite	Al	1	6	3.00	1.935	0.1210	9.5392	1.0114	18.03	-108.0
Hercynite	Al	B	6	3.00	1.937	0.0010	9.5050	1.0125	47.41	-106.0
Ferberite	W	2e	6	6.00	1.938	0.1560	9.3021	1.0339	95.94	-358.5
Al2TiO5 Syn	Al	M2	6	3.00	1.939	0.1260	9.1882	1.0418	128.1	-110.4
Pseudobrookite	Ti	M1	6	4.00	1.941	0.0390	9.4100	1.0239	76.07	-177.5
Al2TiO5 Syn	Ti	M1	6	4.00	1.944	0.1220	8.9351	1.0672	181.8	-175.7
Anatase	Ti	Ti	6	4.00	1.946	0.0140	9.3740	1.0319	113.7	-177.6
Amesite	Al	M3	6	3.00	1.946	0.0270	9.6769	1.0102	33.25	-109.9
Amesite	Al	M11	6	3.00	1.947	0.0220	9.6956	1.0102	33.47	-109.5
Ca-Tschermakite	Al	M1	6	3.00	1.947	0.0660	9.6400	1.0140	44.3	-112.2
Kaolinite	Al	3	6	3.00	1.948	0.0950	9.5337	1.0241	76.79	-85.7
Vesuvianite	Al	Al	6	3.00	1.949	0.0610	9.7539	1.0084	25.81	-104.5
Wolframite	W	2e	6	6.00	1.950	0.1790	9.3978	1.0416	115.9	-358.2
Rutile	Ti	Ti	6	4.00	1.955	0.0170	9.8458	1.0081	28.42	-179.3
Titanite	Ti	Ti	6	4.00	1.959	0.0960	9.9775	1.0052	7.562	-180.4
Brookite	Ti	Ti	6	4.00	1.959	0.0620	9.7409	1.0204	68.63	-178.1
Allanite	Al	M1	6	3.00	1.965	0.0690	10.0526	1.0055	11.91	-99.7
Zoisite	Al	3	6	3.00	1.967	0.1480	9.8664	1.0237	54.86	-110.2
Lepidolite	Li	M2	6	2.26	1.972	0.0180	10.0451	1.0121	41.18	-62.5
Staurolite	Al	3A	6	1.11	1.972	0.1000	10.1247	1.0092	16.57	-27.0
Lepidolite	Li	M1	6	2.24	1.977	0.0510	10.1080	1.0130	41.9	-53.6
Lepidolite	Li	M1	6	2.24	1.977	0.0510	10.1080	1.0130	41.9	-53.6
Clinozoisite	Al	3	6	3.00	1.977	0.1690	10.0086	1.0259	58.06	-110.4
Lepidolite	Li	M1	6	2.26	1.980	0.0170	10.1804	1.0115	39.45	-62.3
Uvarovite	Cr	Y	6	3.00	1.985	0.0000	10.4131	1.0007	2.644	-109.6
Gedrite	Mg	M2	6	2.60	1.987	0.0420	10.3791	1.0058	18.55	-73.6
Eskolaite	Cr	Cr	6	3.00	1.990	0.0280	10.3121	1.0131	45.25	-104.8
Chromite	Cr	B	6	3.00	1.990	0.0010	10.3222	1.0123	46.65	-103.1
Staurolite	Al	3B	6	0.75	1.992	0.1060	10.4412	1.0083	12.26	-16.0
Armalcolite	Ti	M2	6	3.00	1.993	0.1140	10.0136	1.0391	121.5	-170.9
Bixbyite	Mn	M1	6	3.00	1.993	0.0010	10.4222	1.0086	28.65	-108.0

Table 11.4c. Six-fold Sites (continued).

Mineral	Element	Site	CN	Chg	Dist	Sigma	Volume	Q.E.	A.V.	EE(eV)
Magnesiochromite	Cr	B	6	3.00	1.994	0.0010	10.4402	1.0087	32.82	-102.1
Ureyite	Cr	M1	6	3.00	2.001	0.0420	10.5450	1.0090	28.3	-106.0
Epidote 2	Al	M3	6	3.00	2.004	0.1610	10.3952	1.0271	66.33	-106.4
Karelianite	V	V	6	3.00	2.012	0.0540	10.7190	1.0098	32.7	-100.1
Elbaite	Al	Li	6	2.00	2.016	0.0770	10.5238	1.0262	81.23	-49.8
Goethite	Fe	Fe	6	3.00	2.021	0.0820	10.6673	1.0226	69.21	-102.0
Andradite	Fe	Y	6	3.00	2.024	0.0000	11.0462	1.0004	1.357	-106.5
Acmite	Fe	M1	6	3.00	2.025	0.0780	10.8690	1.0130	41.9	-105.1
Tourmaline	Mg	Y	6	2.00	2.025	0.0570	10.7153	1.0230	75.63	-57.7
Hematite	Fe	Fe	6	3.00	2.030	0.0930	10.7541	1.0264	85	-104.1
Ilmenite	Ti	Ti	6	4.00	2.032	0.1190	10.6405	1.0372	116.9	-166.0
Magnesioferrite	Fe	B	6	2.55	2.033	0.0010	11.1510	1.0033	12.21	-77.7
Jacobsite	Fe	B	6	2.92	2.035	0.0010	11.0765	1.0092	34.72	-96.5
Epidote 1	Al	M3	6	3.00	2.036	0.1530	10.8636	1.0283	74.75	-104.7
Pargasite	Mg	M2	6	2.27	2.036	0.0540	11.1352	1.0079	24.34	-72.0
Malayaite	Sn	Sn	6	4.00	2.042	0.0740	11.2996	1.0046	8.453	-170.6
Ulvoespinel	Fe	B	6	3.00	2.044	0.0010	11.2515	1.0084	31.58	-99.5
Bixbyite	Mn	M2	6	3.00	2.044	0.1630	10.2390	1.0790	203.7	-103.8
Pseudobrookite	Fe	M2	6	3.00	2.045	0.2060	10.4678	1.0678	207	-103.9
Pyrophanite	Ti	Ti	6	4.00	2.051	0.1520	11.0674	1.0310	91.38	-165.2
Cassiterite	Sn	Sn	6	4.00	2.053	0.0030	11.2922	1.0145	51.1	-171.5
Lizardite	Mg	Mg	6	2.00	2.053	0.0320	11.2391	1.0182	57.98	-77.1
Gedrite	Mg	M3	6	2.00	2.059	0.0350	11.2735	1.0222	70.81	-56.0
Magnetite	Fe	B	6	2.50	2.059	0.0010	11.6121	1.0015	5.593	-73.8
Schorl	Al	Y	6	2.00	2.060	0.0600	11.2467	1.0252	80.09	-49.0
F-Phlogopite	Mg	M1	6	2.00	2.061	0.0250	11.4748	1.0120	38.66	-57.6
gammaNi2SiO4	Ni	B	6	2.00	2.063	0.0010	11.6627	1.0024	8.431	-50.3
Phlogopite-N	Mg	M2	6	2.00	2.063	0.0170	11.4807	1.0127	41.52	-56.3
Phlogopite	Mg	M1	6	2.00	2.063	0.0250	11.5047	1.0119	38.51	-55.9
Tremolite	Mg	M3	6	2.00	2.064	0.0080	11.4982	1.0135	43.85	-65.9
F-Phlogopite	Mg	M2	6	2.00	2.064	0.0220	11.5150	1.0126	40.88	-57.4
Armalcolite	Fe	M1	6	4.00	2.065	0.1040	10.4180	1.0848	230.8	-54.1
Phlogopite	Mg	M2	6	2.00	2.065	0.0200	11.5227	1.0122	39.71	-55.7
Tremolite	Mg	M3	6	2.00	2.066	0.0070	11.5315	1.0133	43.35	-57.3
Phlogopite-N	Mg	M1	6	2.00	2.066	0.0230	11.5344	1.0129	41.76	-55.0
Clinoenstatite	Mg	M1	6	2.00	2.067	0.0990	11.6140	1.0120	31.8	-48.6
Norbergite	Mg	M3	6	2.00	2.068	0.0750	11.5149	1.0174	56.49	-47.1
Mg beta spinel	Mg	M1	6	2.00	2.069	0.0360	11.7310	1.0050	15.22	-50.8
GammaMg2SiO4	Mg	B	6	2.00	2.070	0.0010	11.7800	1.0026	8.95	-50.1
Gedrite	Mg	M3	6	2.00	2.070	0.0110	11.5946	1.0138	44.75	-54.8
Talc	Mg	1	6	2.00	2.071	0.0150	11.6973	1.0087	28.58	-49.4
Talc	Mg	2	6	2.00	2.071	0.0120	11.6871	1.0086	28.6	-49.4
Tremolite	Mg	M1	6	2.00	2.074	0.0100	11.6961	1.0111	36.73	-65.7

Table 11.4d. Six-fold Sites (continued).

Mineral	Element	Site	CN	Chg	Dist	Sigma	Volume	Q.E.	A.V.	EE(eV)
Tremolite	Mg	M1	6	2.00	2.075	0.0090	11.7230	1.0108	35.75	-55.7
MgSiO3 Ilmenite	Mg	Mg	6	2.00	2.076	0.0950	11.2378	1.0429	143.4	-51.3
Gedrite	Mg	M2	6	2.60	2.076	0.0490	11.7651	1.0094	30.8	-65.1
Diopside	Mg	M1	6	2.00	2.077	0.0310	11.8480	1.0050	17.4	-55.7
Tremolite	Mg	M2	6	2.00	2.077	0.0540	11.8284	1.0073	22.96	-50.1
Chondrodite	Mg	M3	6	2.00	2.078	0.0720	11.6653	1.0179	59.24	-47.5
Liebenbergite	Ni	M1	6	2.00	2.078	0.0260	11.5308	1.0254	90.29	-47.3
Pargasite	Mg	M3	6	2.00	2.078	0.0020	11.5404	1.0240	76.2	-55.3
Orthoenstatite	Mg	M1	6	2.00	2.078	0.0680	11.8250	1.0090	26.5	-53.9
Glaucophane	Mg	M1	6	2.00	2.079	0.0180	11.5769	1.0237	79.39	-53.3
Clinohumite	Mg	M3	6	2.00	2.080	0.0720	11.7030	1.0181	59.46	-48.5
Ilmenite	Fe	Fe	6	2.00	2.081	0.0660	11.4248	1.0354	117.8	-54.4
Gedrite	Mg	M1	6	2.00	2.084	0.0320	11.8768	1.0107	34.66	-49.7
Dolomite	Mg	Mg	6	2.00	2.084	0.0010	12.0598	1.0008	3.075	-51.0
Bunsenite	Ni	Ni	6	2.00	2.084	0.0010	12.0713	1.0000	0.0014	-48.3
Tremolite	Mg	M2	6	2.00	2.084	0.0580	11.9321	1.0081	25.58	-61.3
Mg beta spinel	Mg	M2	6	2.00	2.084	0.0240	11.9663	1.0055	19.38	-50.5
Humite	Mg	M3	6	2.00	2.086	0.0740	11.7797	1.0189	62.53	-47.7
Amesite	Mg	M33	6	2.00	2.086	0.0280	11.6242	1.0277	88.45	-50.7
Amesite	Mg	M2	6	2.00	2.087	0.0370	11.6334	1.0278	88.11	-50.8
Cummingtonite	Mg	M2	6	2.00	2.087	0.0400	11.9501	1.0100	33.24	-43.0
Pargasite	Mg	M1	6	2.00	2.088	0.0260	11.8550	1.0160	50.6	-55.9
Mg beta spinel	Mg	M3	6	2.00	2.089	0.0560	12.0394	1.0072	23.44	-51.7
Cummingtonite	Mg	M3	6	2.00	2.091	0.0160	11.8993	1.0167	54.06	-61.1
Gedrite	Mg	M1	6	2.00	2.093	0.0420	11.9232	1.0171	54.26	-50.0
Glaucophane	Mg	M3	6	2.00	2.094	0.0130	11.7829	1.0262	84.52	-53.7
Forsterite	Mg	M1	6	2.00	2.094	0.0290	11.7711	1.0269	95.28	-46.9
Amesite	Mg	M22	6	2.00	2.096	0.0300	11.7721	1.0286	90.64	-50.4
Amesite	Mg	M1	6	2.00	2.096	0.0210	11.7764	1.0287	91.04	-50.3
Cummingtonite	Mg	M1	6	2.00	2.098	0.0180	12.1099	1.0115	37.69	-51.1
Brucite	Mg	Mg	6	2.00	2.099	0.0010	12.0394	1.0161	52.05	-49.4
Liebenbergite	Ni	M2	6	2.00	2.100	0.0600	11.9660	1.0215	74.89	-51.1
Annite	Fe	M2	6	2.00	2.101	0.0140	12.2102	1.0081	26.86	-51.8
Magnesite	Mg	Mg	6	2.00	2.102	0.0010	12.3599	1.0010	3.548	-46.3
CoOrthopyrox	Co	M1	6	2.00	2.103	0.0560	12.2520	1.0090	26.9	-53.0
gammaCo2SiO4	Co	B	6	2.00	2.103	0.0010	12.3315	1.0041	13.95	-48.8
Norbergite	Mg	M2	6	2.00	2.104	0.1000	12.0294	1.0236	75.64	-47.1
Periclase	Mg	Mg	6	2.00	2.105	0.0010	12.4452	1.0000	0.0021	-47.8
Co beta Spinel	Co	M1	6	2.00	2.106	0.0320	12.3213	1.0077	24.51	-49.1
Chondrodite	Mg	M1	6	2.00	2.107	0.0140	11.9654	1.0277	100.7	-45.8
Clinohumite	Mg	M1c	6	2.00	2.107	0.0270	11.9247	1.0301	107.4	-45.4
Humite	Mg	M1	6	2.00	2.108	0.0230	11.9691	1.0293	105	-45.8

Table 11.4e. Six-fold Sites (continued).

Mineral	Element	Site	CN	Chg	Dist	Sigma	Volume	Q.E.	A.V.	EE(eV)
Clinohumite	Mg	M1n	6	2.00	2.109	0.0240	11.9776	1.0297	106.5	-45.9
Cordierite	Mg	M	6	2.00	2.110	0.0080	11.7984	1.0406	128.4	-48.5
Smithsonite	Zn	Zn	6	2.00	2.111	0.0010	12.5200	1.0008	2.801	-46.0
Lepidolite	Li	M1	6	1.00	2.113	0.0050	12.1471	1.0237	74.59	-16.4
Chondrodite	Mg	M2	6	2.00	2.116	0.0810	12.2446	1.0220	73.96	-48.5
Co beta Spinel	Co	M2	6	2.00	2.117	0.0310	12.5497	1.0051	17.43	-49.5
Clinohumite	Mg	M25	6	2.00	2.119	0.0830	12.2803	1.0230	77.1	-49.2
Co-Olivine	Co	M1	6	2.00	2.119	0.0370	12.1437	1.0294	102.9	-46.1
Annite	Fe	M1	6	2.00	2.121	0.0010	12.5333	1.0095	31.69	-50.7
Co beta Spinel	Co	M3	6	2.00	2.121	0.0470	12.6139	1.0064	21.02	-50.5
Lepidolite	Li	M2	6	1.00	2.122	0.0090	12.2436	1.0270	82.8	-53.6
Lepidolite	Li	M2	6	1.00	2.122	0.0090	12.2436	1.0270	82.8	-53.6
Humite	Mg	M25	6	2.00	2.122	0.0820	12.3449	1.0223	74.91	-48.7
Lepidolite	Li	M2	6	1.00	2.124	0.0260	12.3286	1.0237	74.06	-15.1
Ankerite	Fe	Fe	6	2.00	2.126	0.0010	12.7890	1.0009	3.441	-49.5
Pyrophanite	Mn	Mn	6	2.00	2.127	0.1130	12.3359	1.0289	91.83	-52.9
Forsterite	Mg	M2	6	2.00	2.129	0.0780	12.4009	1.0260	89.49	-50.3
Ferberite	Fe	2f	6	2.00	2.129	0.0580	12.5587	1.0165	56.06	-56.3
Kirschsteinite	Fe	M1	6	2.00	2.130	0.0260	12.1049	1.0427	147.9	-46.1
Hedenbergite	Fe	M1	6	2.00	2.131	0.0520	12.8110	1.0060	17.4	-53.4
Monticellite	Mg	M1	6	2.00	2.134	0.0470	12.4203	1.0287	100.3	-46.8
Orthoferrosilite	Fe	M1	6	2.00	2.135	0.0490	12.8140	1.0090	28.7	-51.8
Zinnwaldite	Fe	M3	6	1480	2.135	0.0190	12.4072	1.0302	99.87	-23.7
Clinohumite	Mg	M26	6	2.00	2.136	0.0830	12.4829	1.0283	97.14	-49.1
Zinnwaldite	Fe	M1	6	1.48	2.136	0.0060	12.4246	1.0299	99.47	-23.6
gammaFe2SiO4	Fe	B	6	2.00	2.137	0.0000	12.9119	1.0051	17.35	-47.8
Humite	Mg	M2	6	2.00	2.137	0.0860	12.4831	1.0291	99.37	-49.2
Clinoferrosilite	Fe	M1	6	2.00	2.137	0.0500	12.8610	1.0080	27.1	-51.8
Co-Olivine	Co	M2	6	2.00	2.142	0.0770	12.6065	1.0269	92.76	-50.0
Siderite	Fe	Fe	6	2.00	2.144	0.0000	13.1224	1.0013	4.589	-45.1
Clinoenstatite	Mg	M2	6	2.00	2.144	0.1310	12.5300	1.0357	109.2	-55.6
Orthoenstatite	Mg	M2	6	2.00	2.151	0.1780	12.4600	1.0489	140.5	-49.2
Wustite	Fe	Fe	6	2.00	2.155	0.0010	13.3512	1.0000	0.0014	-46.7
Allanite	Al	M3	6	3.00	2.157	0.1530	12.6372	1.0433	125.5	-95.6
Fayalite	Fe	M1	6	2.00	2.161	0.0580	12.7370	1.0379	130.1	-45.2
Staurolite	Fe	U2	6	0.08	2.163	0.0490	12.9570	1.0286	85.8	0.3
Staurolite	Fe	U1	6	0.16	2.165	0.0400	12.9660	1.0299	90.4	0.5
Wolframite	Mn	2f	6	2.00	2.176	0.0970	13.2861	1.0246	80.63	-54.7
MnClinopyrox	Mn	M1	6	2.00	2.177	0.0690	13.6020	1.0090	26.8	-50.5
Fayalite	Fe	M2	6	2.00	2.177	0.1100	13.0718	1.0370	124.9	-49.3
CoOrthopyrox	Co	M2	6	2.00	2.182	0.2180	12.8600	1.0600	158.7	-48.5
Gedrite	Na	M4	6	2.00	2.185	0.1380	12.7658	1.0622	202.3	-36.5
Rhodochrosite	Mn	Mn	6	2.00	2.190	0.0010	13.9861	1.0009	3.416	-43.7

Table 11.4f. Six-fold Sites (continued).

Mineral	Element	Site	CN	Chg	Dist	Sigma	Volume	Q.E.	A.V.	EE(eV)
Bustamite	Mn	M3	6	2.00	2.195	0.0360	13.8410	1.0130	40.9	-49.5
Tephroite	Mn	M1	6	2.00	2.206	0.0370	13.4994	1.0398	138.8	-44.0
Spodumene	Li	M2	6	1.00	2.213	0.0830	10.7760	1.2170	549.4	-15.1
Rhodonite	Mn	1	6	2.00	2.218	0.0760	14.1350	1.0210	65.1	-49.2
Pyroxmangite	Mn	2	6	2.00	2.220	0.0920	14.2640	1.0160	52.6	-48.8
Paragonite	Al	M1	6	2.00	2.221	0.0470	13.8510	1.0364	107.9	-99.6
Pyroxmangite	Mn	3	6	2.00	2.222	0.0910	14.3390	1.0140	44.7	-49.3
Pyroxmangite	Mn	4	6	2.00	2.222	0.0910	13.6710	1.0480	163.8	-46.8
Rhodonite	Mn	3	6	2.00	2.223	0.0950	13.5720	1.0540	183.5	-46.8
Pyroxmangite	Mn	1	6	2.00	2.223	0.0790	14.2550	1.0200	61.3	-48.6
Manganosite	Mn	Mn	6	2.00	2.223	0.0010	14.6472	1.0000	0.0014	-45.3
Orthoferrosilite	Fe	M2	6	2.00	2.223	0.2430	13.4300	1.0700	181.2	-47.7
Clinoferrosilite	Fe	M2	6	2.00	2.224	0.2390	13.5530	1.0640	164.3	-47.4
Tephroite	Mn	M2	6	2.00	2.227	0.0880	13.9817	1.0367	127	-47.8
Rhodonite	Mn	2	6	2.00	2.242	0.0950	14.4800	1.0260	88.8	-48.2
Kirschsteinite	Ca	M2	6	2.00	2.268	0.0920	14.5492	1.0468	161.1	-46.0
Avicennite	Tl	2	6	3.00	2.268	0.1620	13.7645	1.0895	224.3	-94.3
Avicennite	Tl	1	6	3.00	2.271	0.0000	15.0460	1.0249	77.81	-93.7
Pyroxmangite	Mn	6	6	2.00	2.272	0.2940	13.8750	1.0980	257	-48.0
MnClinopyrox	Mn	M2	6	2.00	2.281	0.2660	14.4520	1.0750	192	-46.1
Pyroxmangite	Mn	7	6	2.00	2.284	0.2250	14.6310	1.0650	175.9	-46.6
Otavite	Cd	Cd	6	2.00	2.288	0.0000	15.9442	1.0008	2.801	-48.6
Villaumite	Na	Na	6	1.00	2.307	0.0010	16.3712	1.0000	0.0014	-10.9
Pyroxmangite	Mn	5	6	2.00	2.310	0.2770	12.2300	1.2330	455.1	-45.6
Bustamite	Ca	M1	6	2.00	2.348	0.0920	16.0360	1.0510	161.3	-44.9
Ca-Olivine	Ca	M1	6	2.00	2.352	0.0350	15.8955	1.0601	209	-41.0
Calcite	Ca	Ca	6	2.00	2.360	0.0000	17.4683	1.0019	7.092	-46.2
Monticellite	Ca	M2	6	2.00	2.364	0.0800	16.4378	1.0481	165.6	-43.8
Ankerite	Ca	Ca	6	2.00	2.371	0.0010	17.7261	1.0020	7.326	-40.9
Bustamite	Ca	M2	6	2.00	2.372	0.0900	16.2400	1.0640	189.6	-43.1
Wollastonite	Ca	M1	6	2.00	2.373	0.1120	16.4300	1.0580	176.8	-33.2
Wollastonite	Ca	M2	6	2.00	2.381	0.0720	16.6050	1.0560	177.2	-38.6
Dolomite	Ca	Ca	6	2.00	2.381	0.0010	17.9495	1.0016	5.839	-40.6
Glaucophane	Na	M4	6	1.00	2.391	0.0480	12.5551	1.2827	748.6	-10.8
Ca-Olivine	Ca	M2	6	2.00	2.392	0.0570	16.9298	1.0516	180.8	-43.7
Lime	Ca	Ca	6	2.00	2.405	0.0010	18.5532	1.0000	0.0014	-41.9
Soda Niter	Na	Na	6	1.00	2.417	0.0010	18.8280	1.0000	0.0646	-20.4
Tremolite	Ca	M4	6	2.00	2.419	0.1000	12.5358	1.3157	730.3	-35.7
Lawsonite	Ca	Ca	6	2.00	2.421	0.0380	18.3153	1.0225	78.4	-50.7
Anorthite	Ca	000	6	2.00	2.453	0.1190	15.2217	1.1889	513.1	-41.6
Margarite	Ca	Ca	6	2.00	2.455	0.0320	19.6598	1.0021	6.567	-39.7
Gedrite	Na	A	6	0.34	2.550	0.1550	18.7468	1.1201	319.6	-4.4
High Albite	Na	Na	6	1.00	2.597	0.1980	17.9221	1.1990	553.2	-13.4

Table 11.4g. Six-fold Sites (continued).

Mineral	Element	Site	CN	Chg	Dist	Sigma	Volume	Q.E.	A.V.	EE(eV)
Paragonite	Na	Na	6	1.00	2.624	0.1020	23.9471	1.0051	9.37	-16.9
Niter	K	K	6	1.00	2.649	0.0010	24.6829	1.0026	9.565	-18.8
Halite	Na	Na	6	1.00	2.819	0.0010	29.8692	1.0000	0.0021	-8.9
Leucite	K	K	6	1.00	3.015	0.0670	23.9533	1.3256	951.4	-10.3
Sylvite	K	K	6	1.00	3.145	0.0000	41.4961	1.0000	0.0014	-8.0

Table 11.5. Seven-fold Sites.

Mineral	Element	Site	CN	Chg	Dist	Sigma	Volume	E.E.(eV)
Baddeleyite	Zr	Zr	7	4.00	2.159	0.0840	14.5328	-168.9
Gedrite	Na	M4	7	2.00	2.349	0.3740	17.2813	-34.0
Rhodonite	Mn	4	7	2.00	2.356	0.3490	15.8450	-48.1
Rhodonite	Mn	5	7	2.00	2.397	0.3270	18.5670	-46.2
Vesuvianite	Ca	2	7	2.00	2.402	0.0600	18.9250	-37.3
Wollastonite	Ca	M3	7	2.00	2.414	0.1170	20.2910	-82.7
Fluorapatite	Ca	2	7	2.00	2.428	0.1800	21.5056	-43.2
Titanite	Ca	Ca	7	2.00	2.458	0.1300	19.7127	-43.3
Anorthite	Ca	z00	7	2.00	2.490	0.1320	20.6967	-41.5
Malayaite	Ca	Ca	7	2.00	2.490	0.1880	20.6878	-43.7
Anorthite	Ca	0i0	7	2.00	2.502	0.1550	21.0876	-41.3
Anorthite	Ca	zi0	7	2.00	2.533	0.1880	21.9001	-42.3
Monazite	Ce	1	7	3.00	2.534	0.1630	19.6755	-86.3
Zoisite	Ca	2	7	2.00	2.551	0.1780	31.0110	-45.1
Low Albite	Na	Na	7	1.00	2.611	0.2590	22.4571	-14.4

Table 11.6. Eight-fold Sites.

Mineral	Element	Site	CN	Chg	Dist	Sigma	Volume	E.E.(eV)
Hafnon	Hf	Hf	8	4.00	2.188	0.0780	18.7427	-196.0
Zircon	Zr	Zr	8	4.00	2.198	0.0740	19.0044	-168.4
MgSiO3 Perov	Mg	Mg	8	2.00	2.203	0.1710	20.1002	-49.6
Pyrope	Mg	X	8	2.00	2.270	0.0780	20.1374	-48.8
Almandine	Fe	A	8	2.00	2.299	0.0840	20.9299	-47.8
Spessartine	Mn	X	8	2.00	2.326	0.0860	21.6500	-46.8
Fluorite	Ca	Ca	8	2.00	2.364	0.0010	40.6928	-61.7
Uraninite	U	U	8	4.00	2.368	0.0000	40.8764	-147.3
Coffinite	U	U	8	4.00	2.376	0.0570	24.0361	-152.6
Grossular	Ca	X	8	2.00	2.405	0.0910	23.8806	-44.3
Thorite	Th	Th	8	4.00	2.417	0.0530	25.3243	-149.8
Vesuvianite	Ca	1	8	2.00	2.421	0.1050	24.4046	-32.7
Thorianite	Th	Th	8	4.00	2.425	0.0000	43.8969	-143.8
Uvarovite	Ca	X	8	2.00	2.429	0.0750	24.4760	-43.8
Andradite	Ca	X	8	2.00	2.433	0.0720	24.5480	-44.0
Apatite(OH)	Ca	2	8	2.00	2.443	0.1270	26.1375	-43.8
Scheelite	Ca	Ca	8	2.00	2.458	0.0220	26.3758	-49.6
Ca-Tschermakite	Ca	M2	8	2.00	2.460	0.0610	24.5210	-42.3
Gypsum	Ca	Ca	8	2.00	2.463	0.1270	26.1169	-42.6
Jadeite	Na	M2	8	1.00	2.469	0.1690	24.5830	-13.6
Anhydrite	Ca	Ca	8	2.00	2.471	0.0850	26.4332	-44.0
Vesuvianite	Ca	C	8	1.00	2.474	0.2580	32.1330	-4.6
Bustamite	Ca	M4	8	2.00	2.490	0.1200	24.7800	-43.5
Pargasite	Ca	M4	8	2.00	2.490	0.1150	25.8711	-43.7
Ureyite	Na	M2	8	1.00	2.491	0.1710	25.4480	-13.6
Diopside	Ca	M2	8	2.00	2.498	0.1630	25.7600	-41.5
Vesuvianite	Ca	3	8	2.00	2.499	0.0800	27.0015	-50.5
Cordierite	Na	Alk	8	0.00	2.509	0.1600	23.2560	0.0
Hedenbergite	Ca	M2	8	2.00	2.511	0.1770	26.1070	-41.6
SrF2	Sr	Sr	8	2.00	2.511	0.0010	48.7679	-60.6
Tremolite	Ca	M4	8	2.00	2.518	0.1750	26.4916	-43.9
Acmite	Na	M2	8	1.00	2.519	0.1930	26.2950	-13.4
Gehlenite	Ca	Ca	8	2.00	2.563	0.1680	32.4750	-43.0
Melilite	Ca	Ca	8	1.50	2.570	0.1390	32.7450	-38.1
Na-melilite	Ca	Ca	8	1.50	2.572	0.1440	32.8930	-26.2
Akermanite	Ca	Ca	8	2.00	2.577	0.1370	32.9090	-41.3
Chlorapatite	Ca	2	8	2.00	2.577	0.2550	34.1257	-42.3
Clinozoisite	Ca	2	8	2.00	2.579	0.1770	32.8896	-45.7
Epidote 2	Ca	2	8	2.00	2.582	0.1740	32.9768	-45.3
Epidote 1	Ca	2	8	2.00	2.588	0.1740	33.0794	-46.5
Nepheline	Na	Na	8	1.00	2.620	0.1190	26.7431	-12.1
Frankdickson	Ba	Ba	8	2.00	2.683	0.0010	59.4782	-59.5
Microcline	K	K	8	1.00	2.925	0.1110	36.3958	-11.8

Table 11.7. Sites of Coordination Number Greater Than Eight.

Mineral	Element	Site	CN	Chg	Dist	Sigma	Volume	E.E.(eV)
Aragonite	Ca	Ca	9	2.00	2.528	0.0850	37.5424	-41.4
Chlorapatite	Ca	1	9	2.00	2.545	0.1840	31.9771	-42.7
Fluorapatite	Ca	1	9	2.00	2.550	0.1900	31.7503	-42.7
Apatite(OH)	Ca	1	9	2.00	2.555	0.1890	31.9741	-42.0
Zoisite	Ca	1	9	2.00	2.562	0.2230	28.0948	-41.5
Clinozoisite	Ca	1	9	2.00	2.575	0.2500	27.6077	-41.6
Epidote 2	Ca	1	9	2.00	2.578	0.2680	27.4051	-42.2
Epidote 1	Ca	1	9	2.00	2.586	0.2910	27.2489	-41.7
Marialite	Na	Na	9	1.21	2.599	0.2860	32.5482	-18.0
Meionite	Na	Na	9	1.92	2.600	0.1920	32.5975	-39.4
Allanite	Ca	A1	9	2.00	2.613	0.3380	27.5313	-41.4
Strontianite	Sr	Sr	9	2.00	2.636	0.0670	42.4805	-39.3
Elbaite	Na	CaMn	9	1.00	2.671	0.1630	31.4600	-13.8
Tourmaline	Na	Na	9	1.00	2.673	0.1080	31.5639	-14.4
Schorl	Na	CaMn	9	1.00	2.673	0.1080	31.5600	-13.8
Cerussite	Pb	Pb	9	2.00	2.695	0.0510	45.4671	-38.1
Witherite	Ba	Ba	9	2.00	2.807	0.0530	50.8957	-36.5
Celsian	Ba	Ba	9	2.00	2.921	0.1410	41.1651	-35.1
Sanidine	K	K	9	1.00	2.958	0.1330	54.4298	-12.0
Kalsilite	K	K	9	1.00	2.968	0.0240	46.6861	-9.8
Orthoclase	K	K	9	1.00	2.971	0.1330	55.1617	-11.8
Nepheline	K	K	9	1.00	3.016	0.0300	49.9250	-8.2
Perovskite	Ca	Ca	10	2.00	2.647	0.1210	37.1118	-41.2
Celestite	Sr	Sr	10	2.00	2.745	0.1490	45.8961	-41.3
Anglesite	Pb	Pb	10	2.00	2.783	0.1590	47.9685	-40.7
Barite	Ba	Ba	10	2.00	2.879	0.1130	52.9887	-38.7
Muscovite	K	K	10	1.00	3.031	0.2240	51.2668	-12.9
Rb Feldspar	Rb	Rb	10	1.00	3.100	0.0850	67.4802	-10.9
Allanite	La	A2	11	2.00	2.729	0.2480	54.8598	-47.9
Lepidolite	K	K	12	1.00	3.105	0.1720	62.8979	-10.5
Lepidolite	K	K	12	1.00	3.111	0.1690	62.8918	-0.7
Lepidolite	K	K	12	1.00	3.111	0.1690	62.8918	-0.7
Lepidolite	K	K	12	1.00	3.119	0.1530	60.8481	-10.9
Zinnwaldite	K	1	12	1.00	3.126	0.1430	63.8248	-12.1
F-Phlogopite	K	K	12	1.00	3.140	0.1390	77.3672	-7.4
Phlogopite	K	K	14	1.00	3.121	0.1720	80.4488	-10.4
Phlogopite-N	K	K	14	1.00	3.122	0.1760	80.5570	-10.4
Annite	K	K	14	1.00	3.159	0.0680	81.1994	-9.5

REFERENCES

Akimoto, S., T. Nagata, and T. Katsura (1957) The $TiFe_2O_5$–Ti_2FeO_5 solid solution series. *Nature* **179,** 37-38.

Alberti, A., G. Vezzalini, and V. Tazzoli (1981) Thomsonite: a detailed refinement with cross checking by crystal energy calculations. *Zeolites* **1,** 91-97.

Alberti, A. and G. Vezzalini (1983) Thermal behavior of heulandites: a structural study of the dehydration of Nadap heulandite. *Tschermaks Mineralogische und Petrologische Mitteilungen* **31,** 259-270.

Allen, F. (1985) *Chemical and Structural Variations in Vesuvianite* (Ph.D. Thesis), Harvard University, 440 p.

Asbrink, S. and L.-J. Norrby (1970) A refinement of the structure of copper (II) oxide with a discussion of some exceptional e.s.d.'s. *Acta Crystallographica* **B26,** 8-15.

Aurivilius, K. (1956) The crystal structure of mercury (II) oxide. *Acta Crystallographica* **9,** 685-686.

Baur, W. H. (1956) Ueber die Verfeinerung der Kristallstrukturbestimmung einiger Vertreter des Rutiltyps: TiO_2, SnO_2, GeO_2 und MnF_2. *Acta Crystallographica* **9,** 515-520.

Baur, W. H. (1961) Atomabstaende und Bildungswinkel im Brookit, TiO_2. *Acta Crystallographica* **14,** 214-216.

Baur, W. H. and A. A. Khan (1971) Rutile-type compounds IV. SiO_2, GeO_2 and a comparison with other rutile-type compounds. *Acta Crystallographica* **B27,** 2133-2139.

Baur, W. H. (1978) Crystal structure refinement of lawsonite. *American Mineralogist* **63,** 311-315.

Beran, A. and J. Zemann (1977) Refinement and comparison of the crystal structures of a dolomite and of an Fe-rich ankerite. *Tschermasks Mineralogische und Petrologische Mitteilung* **24,** 279-286.

Beyer, H. (1967) Verfeinerung der Kristallstruktur von Tellurit, dem rhombischen TeO_2. *Zeitschrift fuer Kristallographie* **124,** 228-237.

Blake, R. L., R. E. Hessevick, T. Zoltai, and L. W. Finger (1966) Refinement of the hematite structure. *American Mineralogist* **51,** 123-129.

Blount, A. M., I. M. Threadgold, and S. W. Bailey (1969) Refinement of the crystal structure of nacrite. *Clays and Clay Minerals* **17,** 185-194.

Borie, B. (1974) Thermally excited forbidden reflections. *Acta Crystallographica* **A30,** 337-341.

Borodin, V. L., V. I. Lyutin, V. V. Ilyukhin, and N. V. Belov (1979) Isomorphous calcite-otavite series. *Doklady Akademia Nauk SSSR* **245,** 1099-1101.

Bragg, W. L., G. F. Claringbull, and W. L. Taylor (1965) *Crystal Structures of Minerals,* Cornell University Press, Ithaca, 409 p.

Brown, B. E. and S. W. Bailey (1964) The structure of maximum microcline. *Acta Crystallographica* **17,** 1391-1400.

Brown, G. E. (1970) *The crystal chemistry of the olivines (PhD Thesis).* Virginia Polytechnic Institute and State University, Blacksburg, VA. 121 p.

Burnham, C. W. (1967) Ferrosilite. *Carnegie Institution of Washington Yearbook* **65,** 285-290.

Busing, W. R. and H. A. Levy (1958) A single crystal neutron diffraction study of diaspore, $AlO(OH)$. *Acta Crystallographica* **11,** 798-803.

Cameron, M., S. Sueno, C. T. Prewitt, and J. J. Papike (1973) High-temperature crystal chemistry of acmite, diopside, hedenbergite, jadeite, spodumene, and ureyite. *American Mineralogist* **58,** 594-618.

Calligaris, M., G. Nardin, and L Randaccio (1982) Cation-site location in a natural chabazite. *Acta Crystallographica* **B38,** 602-605.

Cannillo, E., F. Mazzi, J. H. Fang, P. D. Robinson, and Y. Ohya (1971) The crystal structure of aenigmatite. *American Mineralogist* **56,** 427-446.

Catlow, C. R. A., J. M. Thomas, S. C. Parker, and D. A. Jefferson (1982) Simulating silicate structures and the stability of pyroxenoids. *Nature* **295,** 658-662.

Clark, J. R., D. E. Appleman, and J. J. Papike (1969) Crystal-chemical characterization of clinopyroxenes based on eight new structure refinements. *Mineralogical Society of America Special Paper* **2,** 31-50.

Cohen, J. P., F. K. Ross, and G. V. Gibbs (1977) An x-ray and neutron diffraction study of hydrous low cordierite. *American Mineralogist* **62,** 67-78.

Cole, W. F. and C. J. Lancucki (1974) A refinement of the crystal structure of gypsum $CaSO_4 2H_2O$. *Acta Crystallographica* **B30,** 921-929.

Colville, A. A. and P. H. Ribbe (1968) The crystal structure of an adularia and a refinement of the structure of orthoclase. *American Mineralogist* **53,** 25-37.

Czaya, R. (1971) Refinement of the structure of $\gamma{-}Ca_2SiO_4$. *Acta Crystallographica* **B27,** 848-849.

DeVilliers, J. P. R. (1971) Crystal structures of aragonite, strontianite, and witherite. *American Mineralogist* **56,** 758-767.

Dollase, W. A. (1968) Refinement and comparison of the structures of zoisite and clinozoisite. *American Mineralogist* **53,** 1882-1898.

Dollase, W. A. (1971) Refinement of the crystal structures of epidote, allanite, and hancockite. *American Mineralogist* **56,** 447-464.

Donnay, G. and R. Barton Jr. (1972) Refinement of the crystal structure of elbaite and the mecahinism of tourmaline solid solution. *Tschermaks Mineralogische und Petrologische Mitteilung* **18,** 273-286.

Effenberger, H., K. Mereiter, and J. Zemann (1981) Crystal structure refinements of magnesite, calcite, rhodochrosite, siderite, smithsonite, and dolomite, with discussion of some aspects of the stereochemistry of calcite type carbonates. *Zeitschrift fuer Kristallographie* **156,** 233-243.

Finger, L. W. (1970) Refinement of the crystal structure of an anthophyllite. *Carnegie Institution of Washington Yearbook* **68,** 283-288.

Fischer, K. (1966) A further refinement of the crystal structure of cummingtonite, $(Mg,Fe)_7(Si_4O_{11})_2(OH)_2$. *American Mineralogist* **51,** 814-818.

Fischer, P. (1967) Neutronenbeugungsuntersuchung der Strukturen von $MgAl_2O_4$- und $ZnAl_2O_4$- Spinellen in Abhaengigkeit von der Vorgeschichte. *Zeitschrift fuer Kristallographie* **124,** 275-302.

Foit, F. F. Jr. and P.E. Rosenberg (1979) The structure of vanadium bearing tourmaline and its implications regarding tourmaline solid solutions. *American Mineralogist* **64,** 788-798.

Foreman, N. and D. R. Peacor (1970) Refinement of the nepheline structure at several temperatures. *Zeitschrift fuer Kristallographie* **132,** 45-70.

Forsyth, J. B., I. G. Hedley, and C. E. Johnson (1968) The magnetic structure and hyperfine field of goethite ($\alpha{-}FeOOH$). *Journal of Physics C* **1,** 179-188.

Robie, R. A., B. S. Hemingway, and J. R. Fisher (1978) Thermodynamic properties of minerals and related substances at 298.15K and 1 bar (10^5 Pascal) pressure and at higher temperatures. *U. S. Geological Survey Bulletin* **1452**.

Robinson, K., G. V. Gibbs, and P. H. Ribbe (1971) Quadratic elongation: a quantitative measure of distortion in coordination poyhedra. *Science* **172,** 567-570.

Robinson, K., G. V. Gibbs, and P. H. Ribbe (1973a) The crystal structures of the humite minerals IV. Clinohumite and titanoclinohumite. *American Mineralogist* **58,** 43-49.

Robinson, K., G. V. Gibbs, P. H. Ribbe, and M. R. Hall (1973b) Cation distribution in three hornblendes. *American Journal of Science* **273A,** 522-535.

Rothbauer, R. (1971) Untersuchung eines 2M1-Muskovits mit Neutronenstrahlen. *Neues Jahrbuch fuer Mineralogie, Monatshefte,* 143-154.

Saalfeld, H. and M. Wedde (1974) Refinement of the crystal structure of gibbsite, $Al(OH)_3$. *Zeitschrift fur Kristallographie* **139,** 129-135.

Sabine, T. M. and S. Hogg (1969) The wurtzite Z parameter for beryllium oxide and zinc oxide. *Acta Crystallographica* **B25,** 2254-2256.

Sahl, K. (1974) Verfeinerung der Kristallstruktur von Cerussit, $PbCO_3$. *Zeitschrift fuer Kristallographie* **139,** 215-222.

Sartori, F., M. Franzini, and S. Merlino (1973) Crystal structure of a 2M2 lepidolite. *Acta Crystallographica* **B29,** 573-578.

Sartori, F. (1976) The crystal structure of a 1M lepidolite. *Tschermaks Mineralogische und Petrologische Mitteilungen* **23,** 65-75.

Sartori, F. (1977) The crystal structure of a 2M1 lepidolite. *Tschermaks Mineralogische und Petrologische Mitteilungen* **24,** 23-37.

Sasaki, S., K. Fujino, Y. Takeuchi, and R. Sadanaga (1980) On the estimation of atomic charges by the x-ray method for some oxides and silicates. *Acta Crystallographica* **A36,** 904-915.

Sasaki, S., C. T. Prewitt, Y. Sato, and E. Ito (1982a) Single-crystal x-ray study of γMg_2SiO_4. *Journal of Geophysical Research* **87,** 7829-7832.

Sasaki, S., Y. Takeuchi, K. Fujino, and S. Akimoto (1982b) Electron-density distributions of three orthopyroxenes, $Mg_2Si_2O_6, Co_2Si_2O_6$, and $Fe_2Si_2O_6$. *Zeitschrift fuer Kristallographie* **158,** 279-297.

Sass, R. L., R. Vidale, and J. Donohue (1957) Interatomic distances and thermal anisotropy in sodium nitrate and calcite. *Acta Crystallographica* **10,** 567-570.

Schramm, V. and K. F. Fischer (1971) Refinement of the crystal structure of laumontite. *Molecular Seive Zeolites-I, Advances in Chemistry Series* **101,** 259-265.

Shannon, R. D. and C. T. Prewitt (1969) Effective ionic radii in oxides and fluorides. *Acta Crystallographica* **B25,** 925-945.

Shannon, R. D. and C. T. Prewitt (1970) Revised values of effective ionic radii. *Acta Crystallographica* **B26,** 1042-1048.

Shannon, R. D. (1976) Revised effective ionic radii and systematic studies of interatomic distancesin halides and chalcogenides. *Acta Crystallographica* **A32,** 751-767.

Shintani, H., S. Sato, and Y. Saito (1975) Electron-density distribution in rutile crystals. *Acta Crystallographica* **B31,** 1981-1982.

Shirane, G., S. J. Pickart, and Y. Ishikawa (1959) Neutron diffraction study of antiferromagnetic $MnTiO_3$ and $NiTiO_3$. *Journal Physical Society of Japan* **14,** 1352-1360.

Simonov, M. A., P. A. Sandomerski, F. K. Egorov-Tesmenko, and N. V. Belov (1977) Crystal structure of Willemite, Zn_2SiO_4, Kristallografiya (Doklady, Akademia Nauk SSSR) **237,** 581-584.

Smith, D. K. and H. W. Newkirk (1965) The crystal structure of baddeleyite (monoclinic ZrO_2) and its relation to the polymorphism of ZrO_2. *Acta Crystallographica* **18,** 983-991.

Smith, J. V. (1968) The crystal structure of staurolite. *American Mineralogist* **53,** 1139-1155.

Smyth, J. R., J. V. Smith, G. Artioli, and A. Kvick (1987) Crystal structures of coesite at 15K and 298K from neutron and x-ray single crystal data: test of bonding models. *Journal of Physical Chemistry* **in press.**

Speer, J. A. and G. V. Gibbs (1976) The crystal structure of synthetic titanite, $CaTiOSiO_4$, and the domain textures of natural titanites. *American Mineralogist* **61,** 238-247.

Speer, J. A. and B. J. Cooper (1982) Crystal structure of synthetic hafnon, $HfSiO_4$, comparison with zircon and the actinide orthosilicates. *American Mineralogist* **67,** 804-808.

Sudarsanan, K. and R. A. Young (1969) Significant precision in crystal structural details: Holly Springs hydroxyapatite. *Acta Crystallographica* **B25,** 1534-1543.

Sudarsanan, K., P. E. Mackie, and R. A. Young (1972) Comparison of synthetic and mineral fluorapatite, $Ca_5(PO_4)_3F$, in crystallographic detail. *Materials Research Bulletin* **7,** 1331-1338.

Suitch, P. R. and R. A. Young (1983) Atom positions in highly ordered kaolinite, *Clays and Clay Minerals* **31,** 357-366.

Svennson, C. (1974) The crystal structure of orthorhombic antimony trioxide, Sb_2O_3. *Acta Crystallographica* **B30,** 458-461.

Svennson, C. (1975) Refinement of the crystal structure of cubic antimony trioxide, Sb_2O_3. *Acta Crystallographica* **B31,** 2016-2018.

Taylor, M. and R. C. Ewing (1978) The crystal structures of the $ThSiO_4$ polymorphs: huttonite and thorite. *Acta Crystallographica* **B34,** 1074-1079.

Tokonami, M., H. Horiuchi, A. Nakano, S. Akimoto, and N. Morimoto (1979) The crystal structure of the pyroxene-type $MnSiO_3$. *Mineralogical Journal* **9,** 424-426.

Uelkue, D. (1967) Untersuchungen zur Kristallstruktur und magnetischen Struktur des Ferberits $FeWO_4$. *Zeitschrift fuer Kristallographie* **124,** 192-219.

Vogel, R. E. and C. P. Kempter (1959) Mathematical technique for the precision determination of lattice constants. *U.S. Atomic Energy Commission* **LA-2317,** 30 p.

Wainwright, J. E. and J. Starkey (1971) A refinement of the structure of anorthite. *Zeitschrift fuer Kristallographie* **133,** 75-84.

Wechsler, B. A., C. T. Prewitt, and J. J. Papike (1976) Chemistry and structure of lunar and synthetic armalcolite. *Earth and Planetary Science Letters* **29,** 91-103.

Wechsler, B. A. and C. T. Prewitt (1984) Crystal structure of ilmenite ($FeTiO_3$) at high temperature and at high pressure. *American Mineralogist* **69,** 176-185.

Weitzel, H. (1976) Kristallstrukturverfeinerung von Wolframiten und Columbiten. *Zeitschrift fuer Kristallographie* **144,** 238-258.

Whittaker, E. J. W. and R. Muntus (1970) Ionic radii for use in geochemistry. *Geochimica et Cosmochimica Acta* **34,** 945-959.

Winter, J. K., F. P. Okamura, and S. Ghose (1979) A high-temperature structural study of high albite, monalbite, and the analbite - monalbite phase transition. *American Mineralogist* **64,** 409-423.

Winter, J. K. and S. Ghose (1979) Thermal expansion and high-temperature crystal chemistry of the Al_2SiO_5 polymorphs. *American Mineralogist* **64,** 573-586.

Wyckoff, R. G. W. (1963) *Crystal Structures.* John Wiley and Sons, New York.

Yagi, T., F. Marumo, and S. Akimoto (1974) Crystal structures of spinel polymorphs of Fe_2SiO_4 and Ni_2SiO_4. *American Mineralogist* **59,** 486-490.

Zachariasen, W. H. (1971) Refined crystal structure of phenacite, Be_2SiO_4. *Kristallografiya* **16**, 1161-1166.

Zemann, J., E. Zobetz, G. Heger, and H. Voellenkle (1979) Strukturbestimmung eines OH-reichen Topases. *Oesterreichische Akademie der Wissenschaften* **6**, 145-147.

Zigan, F. and R. Rothbauer (1967) Neutronenbeugungsmessungen am Brucit. *Neues Jahrbuch fur Mineralogie Monatshefte*, 137-143.

Zoltai, T. and J. H. Stout (1984) *Mineralogy: Concepts and Principles*, Burgess, Minneapolis, 505 p.

MINERAL INDEX